Micropolis
Robotics Primer

AF354514

Micropolis
Robotics Primer

Micropolis Handbooks

Micropolis Press

Micropolis Robotics Primer

from the Micropolis Handbooks series

Micropolis
www.micropolis.com
webmaster@micropolis.com

Printed in the United States of America
and distributed globally by Ingram Content Group, Tennessee, USA
through Lightning Source LLC and Ingram affiliates and subsidiaries.

Cover art, text and book illustrations copyright © 2024 Micropolis Press
an imprint of publisher Clipland GmbH, Westfaelische Str. 64, 47169 Duisburg, Germany, www.clipland.com/corporate, contact@clipland.com.

BISAC book classifications:

TEC037000   TECHNOLOGY & ENGINEERING / Robotics
TEC035000   TECHNOLOGY & ENGINEERING / Reference
TEC004000   TECHNOLOGY & ENGINEERING / Automation
TEC018000   TECHNOLOGY & ENGINEERING / Industrial Technology
TEC020000   TECHNOLOGY & ENGINEERING / Manufacturing
TEC009070   TECHNOLOGY & ENGINEERING / Mechanical

Cataloging Data:

Title: Micropolis Robotics Primer
Author: Micropolis Handbooks
Publisher: Micropolis Press
ISBN: 978-3-9826166-3-6
      pages: 96
      cm: 5.0 mm (0.2in) spine; 216m x 140mm (8.5in x 5.5in)

Disclaimer

The information contained within this book is for educational and entertainment purposes only. Although every precaution has been taken in the preparation of this book, the author and the publisher ("we", "us", "Micropolis", "Clipland GmbH") assume no responsibility for errors or omissions. No warranties of any kind are expressed or implied. The reader agrees that we, under no circumstances, are responsible for any losses, direct or indirect, which are incurred as a result of the use of information contained within this text. Changes are periodically made to the information herein; these changes will be incorporated in new editions of the publication. Readers acknowledge that we are not engaging in the rendering of legal, financial, medical, technical or professional advice. For more information, the author and publisher may be contacted at the previously stated address.
MICROPOLIS PROVIDES THIS PUBLICATION "AS IS" WITHOUT WARRANTY OF ANY KIND, EITHER EXPRESS OR IMPLIED, INCLUDING, BUT NOT LIMITED TO, THE IMPLIED WARRANTIES OF NON-INFRINGEMENT, MERCHANTABILITY OR FITNESS FOR A PARTI-CULAR PURPOSE. Some jurisdictions do not allow disclaimer of express or implied warranties in certain transactions, therefore, this statement may not apply to you.
Statements regarding Micropolis' future direction or intent are subject to change or withdrawal without notice, and represent goals and objectives only.

Trademarks

Many of the designations used by manufacturers and sellers to distinguish their products or services are claimed as trademarks. Where those designations appear in this text and Micropolis and/or the authors were aware of a trademark claim, the designations are mentioned along with their owners and may be additionally marked with a trademark symbol. Their use here within this text is for educational use of the reader and is covered under nominative fair use. Micropolis is in no way suggesting support, sponsorship or endorsement of the owner of these trademarks. Only as much of such marks is used as is necessary to identify the trademark owner, product, or service.

Links

This publication provides links to external content as a convenience and for informational purposes only; they do not constitute an endorsement, support, sponsorship or an approval by us. We bear no responsibility for the accuracy, legality or content of the external site or for that of subsequent links.

Comments welcome

We want our Micropolis Handbooks series of publications to be as helpful as possible. If you encounter errors, omissions or would like to share your comments about this book with us, feel free to contact Micropolis via email or the form found at:

https://www.micropolis.com/contact

Table of Contents

X

XII

Introduction

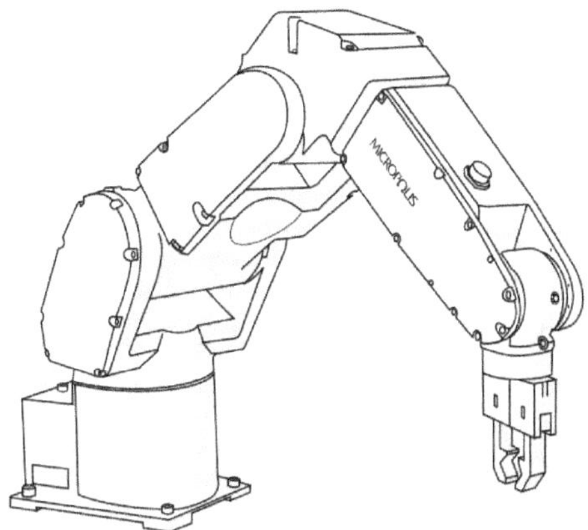

Robotics is an interdisciplinary field, where complex mechanical challenges meet complications in software and hardware design. This text, mainly written in the form of a technical terms glossary, is meant as quick-start guide in mechatronics and robotics engineering. Short paragraphs per term summarize, often complex, matter to provide you with an overview of the field, a first primer on this fascinating topic.

Actuator

An actuator is an essential dynamic component in robotics, designed to generate physical movement or control a mechanism or system by converting energy - from electrical power, compressed air, or compressed fluid - to rotary or linear physical movement. There are three major types of actuators - electric, hydraulic, and pneumatic. Electric actuators, like simple AC or DC motors or more complex servo motors, use electrical power for versatility and precise control. Pneumatic actuators use compressed air to deliver rapid and powerful movements. Hydraulic actuators use pressurized fluid to achieve high force and smooth motion.

Accuracy

"accuracy" in robotics is the measure of a robot's ability reach a pre-set point. It is expressed as the difference between the controller defined setpoint in space and the actual resulting position after the corresponding motion (this is also called "Absolute accuracy"). While Accuracy measures one-time motion accuracy, it influences Repeatability as an error in Accuracy results in cycle-to-cycle variations when a robot is repeatedly instructed to move to a certain setpoint in 3D space (cmp. Repeatability).

AGV

AGV, abbreviation of "Automated Guided Vehicle", is a robotic transport vehicle developed for autonomous material handling and load transportation, or more general a mobile robot platform that strictly adheres to predefined pathways. As such, an AGV is different from an autonomous mobile robot (AMR). AGVs are designed to follow predefined paths by utilizing various sensors and control systems to navigate and interact with the surrounding environment. Commonly deployed in warehouses, distribution centers, and manufacturing facilities, AGVs are essential in logistics and in automation. They operate without an onboard operator or driver. They are battery-powered and programmed to halt their movement if an obstacle blocks their path. The movement continues only when the obstacle is removed. AGVs are equipped with electromagnetic, optical, laser, or other automatic guidance devices, enabling them to travel along predetermined paths while operating safely.

Android

While many people associate the smartphone operating system with the term, an "Android" in robotics is actually a type of robot, although what type of robot it is is not officially defined. In general, through the use in science-fiction, fantasy and everyday culture, an "Android" is a humanoid robot, meaning a robot that isn't only emulating the human form in shape but is also often made of flesh-like material. The distinction is usually, that a (humanoid) Robot is mechanic or metallic in appearance, a more engineered technical looking structure, while an Android appears to be a real organic being, through the use of softer and lifelike looking materials. Although the term "Android" (from Ancient Greek "andro" meaning "man" or "male") itself denotes a male looking robot ("robot boy" or or "robot man"), an Android may just as well be a female-looking humanoid robot. That said, Androids with a female appearance ("robot girl" or "robot woman") may also be called "Gynoids". The abbreviated form of Android, a "Droid", was popularized mainly through the use in George Lucas' Star Wars franchise and has seen more frequent use since then in many areas. The term "Cyborg" is again different from "Android" as a Cyborg is a being that is a combination of organic and mechanical parts ("bionic").

Animatronics

a portmanteau of "animate" and "electronics", Animatronic is a term used to describe machines that are used as puppet or automatons, driving creatures, characters and animals, usually in theme parks. The term has been made popular by Walt Disney's imagineering team WED which started to use these machines for characters in the Disneyland theme parks. Over the years, the animatronic mechatronics have seen vast improvements, adding audio and working on the optimization of human-like or life-like movements under 24/7 operation. Around 2015, Walt Disney Imagineers built upon these techniques and introduced "Automatronics", adding a mobility platform to Animatronics, allowing robot characters to move through their environment and interact with guests.

Arduino

Popular microcontroller platform and ecosystem for DIY robots, or DIY "maker projects" in general. Consisting of open-sourced software and hardware components, the Arduino ecosystem, as a platform, is designed as a building block for physical computing projects and use in education. Microcontroller programming is done in a C dialect, with an easy-to-use IDE and a rich code module ecosystem.

Articulated robots

are robots with rotary joints (vs. linear or cartesian robots with linear actuators along a linear axis). In articulated robots, the joints are usually connected in a chain-like "one joint connected to another"-structure. The number of joints isn't important for the notion of being an articulated robot. Industrial robots usually feature 3-6 articulated axes.

Artificial intelligence (AI)

A.I. is a multi-disciplinary field of computer science that studies and develops technologies used in machines to simulate (emulate) human intelligence in making decisions and performing tasks. The main focus of AI is the development of cognitive skills like natural language processing (NLP), speech recognition (SR), machine vision (MV), learning, reasoning, self-correction, problem-solving, and creativity. Today, early implementations of artificial intelligence have found application in automating tasks such as customer service, fraud detection, lead generation and quality control. It has been empirically confirmed that artificial intelligence can perform some specific tasks faster and better than humans, particularly repetitive and detail-oriented tasks. Artificial intelligence is commonly categorized into weak (narrow AI) or strong (artificial general intelligence, "AGI", wikipedia [en.wikipedia.org/wiki/Artificial_general_intelligence]).

Automated delivery system

in last-mile logistics scenarios, one task delegated to robots is mail and parcel delivery. E-Commerce is gaining increasing market shares and customers not only shop goods, but groceries and food as well. Delivery bots are mobile platforms, able to navigate neighborhoods and deliver parcels or goods to a customer's doorstep. While UAVs (unmanned aerial vehicles) delivering packages are being considered ("delivery drones"), a wheel-based Autonomous Mobile Robot (AMR) has efficiency advantages in dense or easily reachable urbanization and neighborhoods. Ground-based automated delivery robots usually are four or six wheeled and resemble the "Mars rover" shape. Usually, the most part of their structure is a large compartment for payload. Some systems feature lift or delivery systems to present their contents to customers, other than just opening a lid and expecting the customer to take out contained goods.

a common design for urban autonomous delivery robot rovers, most of the structure is payload compartment

Autonomous Mobile Robot (AMR)

An Autonomous Mobile Robot (AMR) is a robotic system developed to operate in dynamic environments, real world applications. AMRs use clever algorithms, sensors, machine vision and sometimes artificial intelligence (to a certain degree, as of 2024) to adapt to changes in their surroundings and find new paths to reach their (sometimes predefined) destination. AMRs and AGVs operate with fundamentally different technologies, including perception and navigation software, resulting in different capabilities and potential applications. AMRs provide flexibility, adaptability, and autonomy, which makes them well-suited for various industries and applications, e.g. as people-mover, in logistics, manufacturing or in agriculture. An AMR in logistics could operate in warehouses, without fixed paths or definition of predefined routes. AMRs then could operate intelligently and swarm-like, automagically optimizing pathways for example and this way improve overall productivity. A "less autonomous" system, which is mostly teleoperated, is usually labeled as an "Unmanned Ground Vehicle" (UGV).

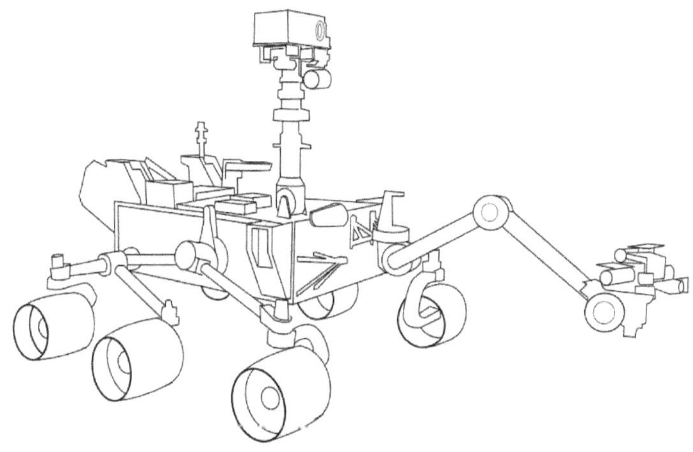

Example of a robotic rover, NASA's 'Curiosity', as deployed on Mars, partly autonomous and teleoperated from Earth

Autonomous

In robotics, the term "autonomous" describes a robot's ability to make decisions independently and perform tasks without direct human intervention, relying on various computations of its systems.

Autonomous vehicle

An autonomous vehicle (AV) refers to a vehicle capable of sensing its surrounding environment and performing operations without a human operator. AVs use various sensors - cameras, radar, lidar, and GPS - in combination with clever software algorithms (artificial intelligence) to control their movement and travel between designated destinations. An AV can just as well be described as an Autonomous Mobile Robot (AMR).

AUV

an "Autonomous Underwater Vehicle" (AUV) is a robotic system designed to operate independently in underwater environments, performing tasks without a human operator. These vehicles deploy sensors, global positioning systems, and acoustic positioning systems for navigation and sensing of surrounding environment. Their propulsion mechanisms provide autonomous movement, making AUVs useful in oceanography, marine biology, and environmental monitoring, particularly for tasks which are challenging or impossible for humans to fulfill. A mostly teleoperated underwater vehicle would be an "Unmanned Underwater Vehicle" (UUV).

Axis naming conventions

With robot articulated arms, axes are usually named in a consistent way, so that technical documentation or technicians may refer to the individual axes equally consistent. In general, a robot's axes are usually named numerically ascending, from the base or root element onward. A "rotary base" is axis one, the first pivot point sitting on top of that base "axis 2" and so forth. Some manufacturers use letters - "A", "B", "C", ... - in alphabetically ascending order. Other manufacturers have defined specific letters for specific axes, for example Yaskawa robots names the rotary base the "S" axis ("Swing axis"), "L"-axis ("Lower arm") moves the robot body back and forth, "T"-axis ("wrist Turning") is the end effector's wrist rotation, etc. With cartesian robots, axes are usually named according to the cartesian coordinate system, "X", "Y" and "Z" being the up/down movement axis. With articulated robot arms, it is common to use a human analogy in the naming of robotic links and joints, speaking of the "shoulder joint axis", the "elbow joint axis" and the "wrist joint axis".

Backlash

usually unwanted (thus challenging) trait of robot arms, actuators and gearboxes where a movement system shows flexibility against physical outside force against its mechanic. Backlash reduces overall accuracy of robots and needs sophisticated systems to be mitigated, usually in hardware and kinematic software. Backlash, meaning flexibility, is an inherent trait in humans and animals, and leads to numerous advantages over the rigid design of current robotic systems in daily tasks while having downsides in repeated movement precision as is important in industrial automated manufacturing applications. As part of Bionics, robot developers often try to incorporate flexibility and adaptability into their designs to improve overall real-world performance of robots.

Bevel gear

a "Bevel gear" assembly is a type of gear train where rotation is transmitted around angles of 90 degrees or less. In addition, bevel gears can shift torque. Universal joints are similar, yet provide no torque shifting. See Hypoid gearbox for an illustration.

Bionics

is a term coined by Jack E. Steele in the 1950s and is used to describe engineering where nature and biology is used as inspiration for technical solutions. Related terms are "biomimetics" and "biomimicry". While some technical solutions end up in a mechanism resembling nature by accident, in bionics the approach is usually to look into nature's problem solving first. A slightly different use of the term "bionic" is when the word is used to describe a combination of a living organism with mechanical or robotic components. While the term "cyborg" is usually more appropriate for such combinations, the bionic aspect of the technical components of such a being - so that these mechanic components interact well with the biological organism - is then used for the whole person. An example of such a generalized use would be the 1970s US television series "The Six Million Dollar Man", centering on a "bionic man" or the spin-off series "The Bionic Woman".

Brushless Motor

a "brushless DC electric motor" (BLDC), sometimes also "electronically commutated motor" is a electrical motor where an electronic controller switches current to a number of electromagnetic coils to induce a rotation into the rotator element. A brushless motor may have an outside in or inside out design (an "outrunner" places the stator with its control magnets at the center, making the rotator on the outside spin around the central static element. In comparison to a "Brushed DC electric motor", the brushless motor eliminates the physical contact between stator and rotator, reducing noise, coal and metal abrasion, which leads to longer component lifetime (less maintenance) and easier implementation in clean-room environments (less/no encapsulation).

What is a cable?

A cable is a collection of wires. While the terms cable and wire are often used interchangeably, there's actually a difference. The wire is the conductive metal part, often copper, aluminum or an alloy, which may be made of a solid extruded piece of metal or multiple strands. A wire doesn't necessarily have to be protected on the outside by a non-conductive material. However, a cable is a collection of wires where each wire is protected (insulated) against contact with the other wires in this cable by some form of non-conductive material, individual thermoplastic sheaths around each wire or by embedding all individual wires in some common insulator material. The collection of wires inside a cable is usually wrapped together by some form of outer cover, made from the same or a different insulator material as used for the internal wires. It is common to have wires and/or cables being wrapped by silicone, rubber, textile meshes, thermoplastics, vinyl, rubber, polyurethane or heat-shrink plastic. With many cables, complex cabling or in combination with movement, cable and wiring systems may need some form of "Cable management".

What is a Cable harness?

Often when cables are bound or assembled together as part of a thicker bundle of cables, some sort of "Cable management" is used to improve maintainability, operational safety and reduce faults, by preventing wiring failures or people or equipment getting entangled in cables. Some simple form of cable management is pulling individual cables through cable ducts, but during manufacturing, when the same wiring setup is repeated many times, this may lead to wiring faults or unnecessary stress on individual cables. Thus, in the early 20th century, it became popular to bundle cables in ad-hoc bound or pre-assembled cable harnesses. Sometimes different terms are used to describe different types of Cable harnesses ("wire harness", "wiring harness", "cable assembly", "wiring assembly" or "wiring loom"). A "wire harness" is usually the most simple form of a cable assembly. Here, a bundle of usually single core wires (so its probably better to speak of a "wire harness") is assembled into one bundle of wires, each isolated against each other but bundled. Over the years different techniques of "cable lacing" were used, where rope, tape, cable ties or similar are use to knit cables together, sometimes in elaborate patterns. A wire harness usually has each single wire visible.

More complex are "Cable assemblies", where a number of cables are combined into one (thicker) sleeve. In contrary to a wire harness, a Cable assembly usually is a bundle of cables, meaning the individual "wires" are actually multicore cables. A cable assembly has all included cables protected by a single outer protective sleeve, giving the appearance of one thicker cable.

A "Cable loom" (sometimes less exact "wiring loom", or "electrical loom") in turn is a bundle of multiple Cable assemblies, wires and combinations thereof, into one complex system.

Many cable looms have the appearance of wiring harnesses with the individual "strands" visible, but with cable looms, each component is either a multicore cable or wire connection. Wire harnesses and Cable looms often are pre-fabricated on dedicated pinboards, where a defined routing layout is used to route wires and cables into a specific pattern, giving the resulting assembly a specific shape or layout, where individual wires or cables may have different lengths, following defined paths or breaking out from the harness/loom at specified points. After a build, technicians can tests such pre-assembled harnesses/looms by connecting the assembly to test circuits. During installation of such assemblies, having a tailored bundle decreases installation time, prevents mistakes and helps with the standardization of work processes.

Cable Management

Cable management as a general practice is the process or guideline of laying, routing and pulling cables in a tidy and organized way. With moving systems, where cabling is applied outside the moving structure, like with robots or when connected modules move relatively to each other, carefully controlling cable flex improves security and maximizes cable life. Thus, cable carriers for cables and hoses are commonly used on industrial robots or in rack equipment [www.micropolis.com/support/kb/rack-mounting-faq] to optimize cable management. From a more general perspective, cable management is an important aspect of mechanical and electrical engineering, inside machines and housings and cabinets. One important technique is to gather a number of cables in "Cable ducts" or assemble multiple cables as part of a "Cable Harness".

Cage

Security assembly in the form of a safety box ("cage") around a robot, to prevent humans or other equipment to enter the work envelope of a robot and get hurt or damaged.

CAN Bus

short for "Controller Area Network", a serial data bus for industrial digital communication networks. Originally developed for on-board vehicle electronics and device control, it is popular in other areas of industrial automation. Developed by Bosch in 1983 and presented with Intel in 1986/87 its design purpose was lowering cost and complexity in cars through reduction of cable looms in automobiles.

Cartesian parallel manipulators

is an interesting robotic implementation where parallel-connected kinematic linkages (parallel not meaning geometrically parallel) are used to hold a platform. Each linkage in turn in mounted on a linear actuator, representing one of the three cartesian axes. Controlling the actuators in unison then enables the system to move the central platform in 3D space.

Cartesian robots

also "Cartesian coordinate robot", "X-Y-Z" robot or "linear robot", is a robotic structure where each of the three cartesian axes is manipulated linearly by actuators, vs. rotational as common in robotic arms. Cartesian robots sometimes implement movement on two axes, as for example in XY Gantry systems, computer plotters, CNC machines, Computer storage Tape Libraries or Warehouse automation. Other machinery, often implementing the third cartesian axis as a linearly driven piston, are Gantry-Robots (cmp. XY-Gantry) where a mechanical gripper either can be moved up and down or sideways (as in automated computer storage Tape Libraries, for example).

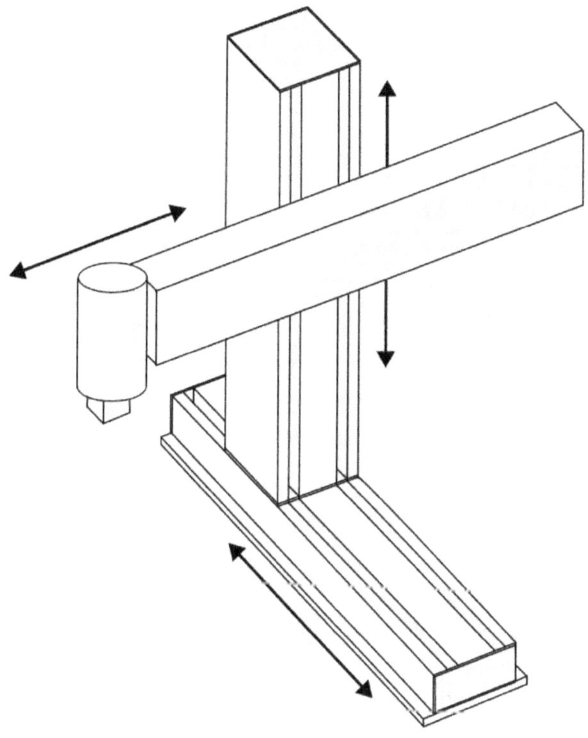

One implementation of a cartesian robot, with three linear tracks

Centering Station

Traditional industrial robots have only limited capabilities to adapt to positional change of handled objects. In automation, robots rely on the environment within their work envelope to remain unchanged so their programmed movements lead to predicted results. When workpieces enter a robot's security cage or work envelope, such workpieces need to be aligned to the exact position of where the robot expects the object as part of its movement cycle. A centering station is a piece of mechanical equipment that does just that: align a workpiece for a robot according to predefined coordinates. This kind of centering is needed when roughly forged objects, (die) casting intermediates or blanks need to be handled by a robot for further processing. As such, a centering station can be regarded as the other end of a homing routine (cmp. "Homing Routine" and "Singling Station"). Centering can be achieved through a number of technologies. Common solutions are simple contour-based limit stops, threading holes on specific clamping jaws or locating pins, or more elaborate active techniques, like optical scanning and algorithmic re-alignment of blanks or workpieces (or a robots waypoints) for subsequent processes.

Closed Loop Control

Robots are usually controlled by closed loop systems, in "closed-loop control". The term describes a system where controller outputs to actuators are measured via sensors that feed their values back into the controller to align setpoints with real-world effects (sensor feedback). Compare "Feedback Loop".

Coaxial shafts

In robots, oftentimes multiple different rotary movements have to be applied to concentric gears to enable the rotation of a robot's wrists around multiple axes. Usually this is implemented via a set of stacked belt-driven rollers, transferring their force onto coaxial shafts.

Collaborative Robot (CoBot)

Short for the collaborative robot, a cobot is an industrial robot developed for operating in a shared workspace with humans. Designed for assistance and collaboration, mainly in the manufacturing industry, they are programmed to protect the safety of their human coworkers. Cobots are intended to enhance productivity by performing various tasks such as parts assembly, packaging automation, material handling, machine operation, product quality control and collaborative research. They are usually designed to be easily programmable and safe to operate.

Market share of Collaborative Robots suppliers

1. **Universal Robots**
   industry wide recognized Danish manufacturer of industrial collaborative robot arms. As a first mover, the supplier captured nearly 50% of the cobot market in 2017, only 9 years after introduction of **UR5** in 2008. Website [www.universal-robots.com/]

2. **FANUC**
   the world-leader in industrial robots is second in the cobots market (about 10%), offering robots with a sen-

sitive shell and dedicated lighter cobot designs. Website [www.fanuc.co.jp/]

3. **TechMan**

    with a market footprint similar to FANUC (~ 10%), laptop manufacturer Quanta Group has established itself among cobot suppliers. Website [www.tm-robot.com/en/]

4. **RethinkRobotics**

    about 8% market share in 2017 is the result of RethinkRobotics bar-tender-like Baxter and Sawyer cobots. Website [www.rethinkrobotics.com/de/]

5. **AUBO**

    like Rethink, AUBO from China can check off an 8% market share in 2017. Website [www.aubo-cobot.com/]

Data Source: Robotics Business Review, 2017 [www.roboticsbusinessreview.com/manufacturing/cobot-market-outlook-strong/],

Collision detection

Collision Detection is an advanced robot control feature implemented by some industrial robot systems and usually all robots operating in an environment shared with humans. The idea is to predict and avoid any collision with surrounding structures or people. Collision detection uses sensor data to predict potential collisions and identify intersection points. The purpose of collision detection in general is to prevent unwanted physical interactions, avoid damage and ensure the safety of robots and their surroundings.

Controller

The controller is a crucial component in robotics designed to command, direct, and regulate a robot's behavior. This mechanism, commonly known as "the brain" of the robot, receives electrical input signals from various sensors, external motion axes, vision systems, etc. It processes these electrical signals based on programmed instructions and sends output signals to actuators to perform the desired action.

Conveyor

A conveyor is a transport device used in robotics to handle materials and automate the transport process. In robotics, conveyors are usually integrated with robots as part of a work cell to enhance the automation of various operations. They are responsible for the synchronized transfer of materials to robots and their further transfer when the operation is finished. Different conveyor types, such as belt conveyors, roller conveyors, and chain conveyors, are widely used in manufacturing, warehouses and distribution centers. They can operate mounted on the floor or as overhead conveyors.

Counterweight system

A coil or similar loaded mechanism to counterweight a robotic arm's weight in order to lessen the stress on electric (or similar) servo drive systems.

Cybernetics

broad concept of circular causality (feedback) in a wide range of topics. In robotics, it is often used in combination with neural networks, medical cybernetics or the "social machine" (machine and human working together, "Industry 5.0").

Cylindrical Robots

Cylindrical Robots, also "Cylindrical coordinate robots" are a form of robot with a circular, cylindrically shaped work envelope. Usually, the design is a rotary base with a vertical tower. A horizontal arm is able to move linearly vertically on this tower. The horizontal arm can either be move sideways or the end effector can move along the arm. In sum, this setup allows the robot to reach points within a cylindrical area minus the dead-zone of the robot's base pedestal. Cylindrical robots are usually used to mediate between two or more machines, conveyor belts or work stations. Compare "Robot Types".

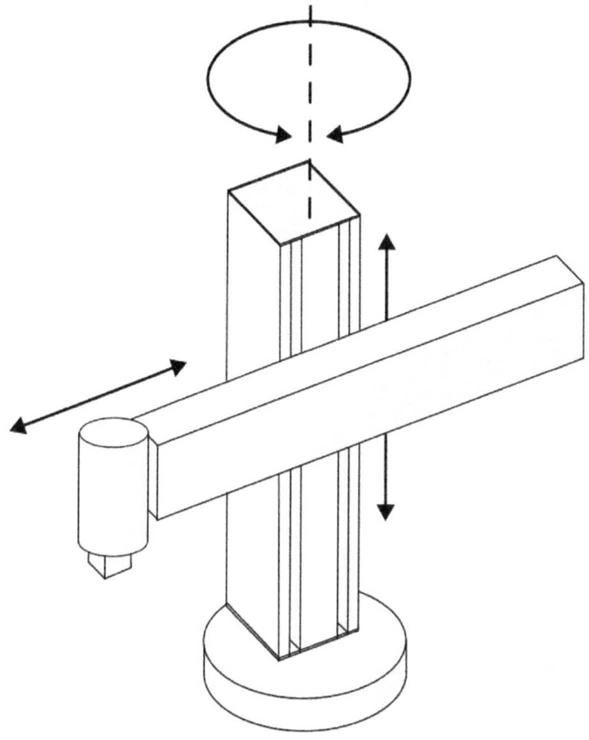

a Cylindrical robot offers rotational movement of its main column, up and down as well as sideways (in and out) movement of its end effector

Dead Zone

Volumetric area inside a robot's work envelope that is not reachable by the robot due to its physical design and constraints. With cylindrical robots, for example, the dead zone is the area of the robot's base pedestal. Cartesian robots usually have no dead zone inside their work envelope as all supporting robot structure is outside the work envelope.

Degrees of Freedom (DoF)

is the number of axes a robot is able move, turn or modify. Bending a joint is one degree of freedom, for example. In sum, the DoFs a robot has describes a robot's flexibility and movement capability, including rotational and translational (linear) movements. Thus, robots with greater degree of freedom, or more degrees of freedom, are more flexible and adaptable.

How many degrees of freedom does a human arm have?

Not considering finger movement, the human arm has seven principal degrees of freedom.

1. Shoulder extension and flexion (rotating the upper arm around the shoulder, to the front and back of the shoulders/body, as seen from the front
2. Shoulder abduction and adduction(seen from the front, moving the shoulder left and right, or the arm away from the body and closer to the body)
3. Shoulder lateral and medial rotation (twisting the upper arm, rotating the lower arm and wrist with it)
4. Elbow extension and flexion (bending and stretching at the elbows)
5. Forearm pronation and supination (twisting the forearm around its axis)
6. Wrist abduction and adduction (moving the wrists sideways, like when waving the hand)
7. Wrist extension and flexion (bending at the wrists, down and upwards, towards the palm of the hand and away)

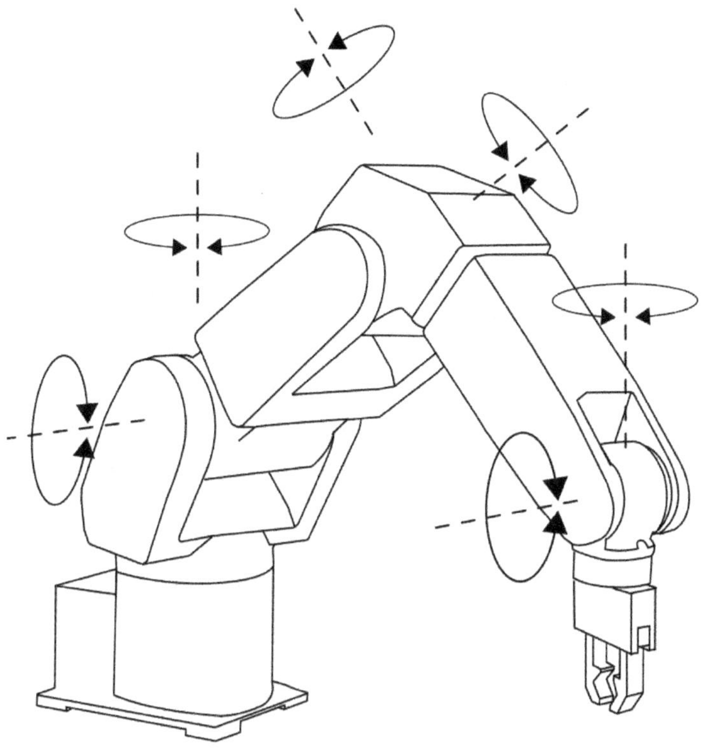

Six Degrees of Freedom (6-DoF) of a jointed articulated robot arm

Delta Drive

Patented drive mechanism invented for Trumpf sheet metal NCT punching machines used in automation, where two parallel leadscrew drives move the punching turret. When both screws turn in the same direction, the turret moves side to side-to-side. When the leadscrews turn in contrary directions, a linear force towards the turret is exerted and presses against angled sliding brackets, resulting in a downward motion to apply

a vertical force, to punch forms into the workpiece. Video
[www.youtube.com/watch?v=cDYhRYp0Gng]

Delta Robot

a Delta Robot (sometimes "parallel robot") is a type of parallel
robot that uses three arms (forming a delta, or triangle). The
arms are constructed to use parallelograms which results in a
fixed orientation of the end effector mount. In order to move
the robot, the arms are connected to three rotary actuators
through a servo arm. Some Delta Robots use four arms instead
of only three. Similarly to what "CoreXY" (cmp.) does for
cartesian gantry systems, the placement of the (heavy) actuators
in the robots base while the robots arms are mere (light) rods,
the actual moved mass of the end effector is reduced and the robot can operate at very high speeds. This makes Delta Robots
popular in fast pick and place applications, e.g. in electronics or
packaging of small goods. The delta robot is related to "Cartesian parallel manipulators".

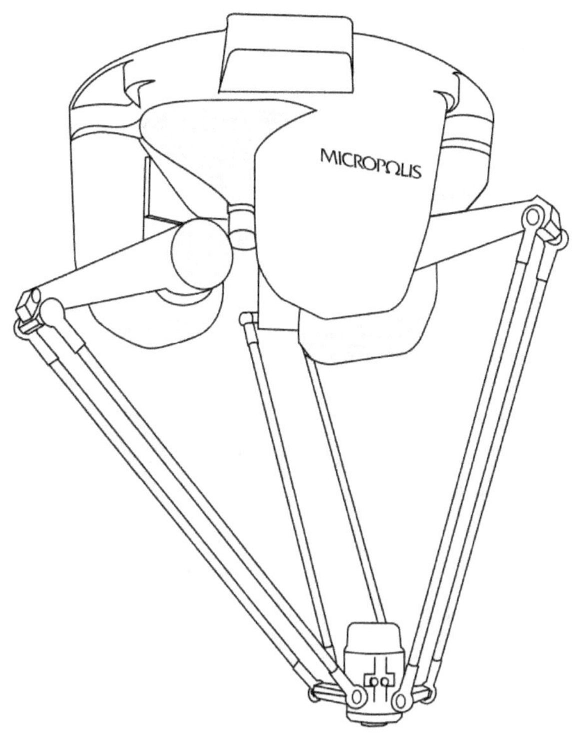

Illustration of a delta robot as used in fast-paced pick and place automation applications

Digital Twin

with the proliferation of IoT, industry 4.0 and the digitization of processes, the term digital twin (also "factory double") began to emerge for complex representations of real-world objects in a digital realm, as a digital simulated counterpart. Like Computer Aided Design (CAD) helped reduce iterations in development, a digital twin of robot or automation setups helps to reduce the installation time while avoiding planning mistakes. During operation, a digital twin can alleviate the burden of machine and robot maintenance, help find and implement optimizations and enables operators to step-in before operational disruptions occur. In combination with service contracts and remote monitoring by robot manufacturers, a digital twin may reduce downtime, improve reliability and offers a predictable cost of ownership of robotic equipment.

Direct Drive

a "direct-drive mechanism" is any mechanical design where the force or torque from a prime mover (a motor) is transmitted directly to the effector device (wheels, robot joint) without involving any intermediate couplings (transmission, gearbox, gear train, belt). In recent years, many mechanical designs have begun to use direct drive as it became possible through advancements in electric motor design and control. Direct drive usually requires a motor to be able to emit large torque through lower rotational frequencies, for example an Electric Vehicle (EV) needs to accelerate from halt to drive - without transmission a direct drive motor has to be able to overcome the initial higher resting force of the assembly. Direct drive robot actuators based on brushless motors are emerging (as of 2024). Direct drive is usually quieter as high ratio gearsets use parts rotating at high frequency and with direct drive these parts can be eliminated.

Direct teaching

Direct teaching is the process of inputting instructions into the robot's memory, "teaching" it to execute specific operations.

This process is accomplished by using low-level and high-level programming, teaching pendants, or manual control.

Differential

is a mechanical device that either distributes one incoming force on two outgoing shafts, or combines the force of two incoming shafts into one motion. It can be embodied as a fixed gear train or with a pulley system of differential belts. In some robot arms, a differential operated by two motors, either moves an arm around one axis when both motors are running synchronously into the same direction, or the arm is rotated around another axis when the motors turn in contrary directions.

Differential wheeled robot

is a type of mobile robot that uses two (separately) driven wheels placed on either side of the robot's body for locomotion. In order to stabilize the robot, a third support element is usually used, to form a triangular ground shape, like a simple thorn or (undriven) caster wheel.

> Related reading:
> - Thesis paper on a differential steering system of robot wheel actuators [rossum.sourceforge.net/papers/DiffSteer/DiffSteer.html]

Domestic robot

a long standing utopian dream is a service robot capable of freeing the modern civilized society from mundane household chores. A household robot (or "home robot") could be the answer and is a popular trope in science-fiction, especially during the American 1950s and 1960s. A home robot could wash the dishes, take the (robot) dog for a walk, work in the kitchen, prepare dinner etc. A domestic robot fulfills the role of a butler or housemaid, is a personal robot, solely meant to help its owner in a mostly private environment. Some domestic robots are meant for indoor use only. Many indoor home robots are toy robots, programmable for fun and education. Domestic outdoor robots are common garden helpers, like robotic lawn mowers, pool cleaning robots or robots for window cleaning. All these applications share that the status quo (as of 2024) is far from what domestic robots were dreamed up to be in the early 20th century.

Drone

the term "drone" might refer to any kind of unmanned vehicle, although (as of 2024) it is common to refer to "unmanned aerial vehicles" as "drones". Compare "UAV" for more about aerial drones. It is somewhat common to speak of drones when referring to teleoperated unmanned marine vessels. Drones are also differentiated by the amount of autonomy a system is capable of. Mostly teleoperated drones operating on land are usually described as "Unmanned Ground Vehicles" (UGVs) with a transition to designate a platform as "Autonomous Mobile Robots" (AMR) depending on its autonomy.

End of Arm Tooling (EOAT)

service by automation vendors to construct, machine or customize grippers, suction cups and other end effector attachments to specific automation needs.

Emergence

from Latin "emergere" (to emerge, to surface, to arise), is when a system has or develops properties which are "more than the sum of its parts", which the parts don't have on their own.

End effector

An end effector is a peripheral device attached to the wrist at the end of a robot's arm, designed to perform a particular operation. It usually serves as a mechanical or electromechanical component within a robotic system that can be categorized into three main types - process tools, grippers and sensors.

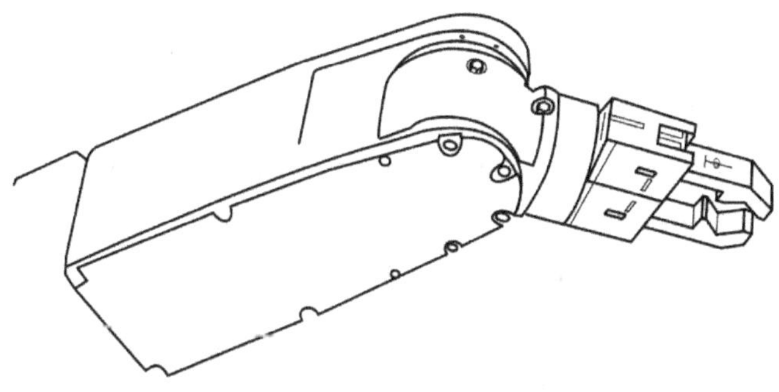

Robot parallel gripper, end effector

ESD protection

ESD is short for "Electrostatic discharge". In robotic applications, it is very common to apply measures to protect the robot system against uncontrolled electrostatic charging or discharging as an ESD could damage things being handled by the robots. For example, safe handling of sensitive electronic components by a robot is only possible with ESD protection.

EtherCAT

short for "Ethernet for Control Automation Technology" is a real-time optimized Ethernet standard, open-sourced as IEC-Standard 61158 and headed by Beckhoff Automation. Coordination among users of the standard is done through the EtherCAT Technology Group (ETG). Technology patents protect the IP of the standard and require exclusive licensing through Beckhoff for implementation.

Exteroception

Exteroception (from Latin "extra", external or outside) is a robot's ability to sense its external surroundings, including the robot itself. Robots gather information about the external environment by processing the data received from various sensors, like - sonars, lidars, cameras, microphones, infrared sensors, etc. Unlike exteroception, proprioception represents the robot's awareness of its own internal environment - its internal state, like joint angles or motor positions. Together, proprioceptive and exteroceptive sensors form the perceptual system of a robot.

What was the First Robot?

This question can only be answered when the concept of what a robot is is well defined. Early non-human agents were known in ancient Greek and similarly old mythologies. They were described as being mechanical puppets, automata made from metal or even gold. Some were described as being similar to mechanical clocks while others were driven by magic or mystic forces. For example, the "mechanical knight" designed by Leonardo da Vinci (cmp. Humanoid Robots) was actually an automaton, not a robot. It was not until the beginning of the 20th century that mankind developed the necessary engineering skills to fabricate electrically driven motors and, with it, dream up machines, humanoid robots, that were the culmination of those skills. Many robots were inventions of fiction of this time. "Tinman" from L. Frank Baum's "The Wonderful Wizard of Oz" (published in 1900) is an often cited first robot. The actual term "robot" originated in "R.U.R." (abbreviation of "Rossumovi Univerzální Roboti", meaning "Rossum's Universal Robots"), a 1920 theater play by Czech author Karel Čapek. In 1927 Fritz Lang premiered his early cinema masterpiece "Metropolis" and with it "Maria" an all-metal female (Gynoid) robot revenant of a love interest of one of the play's characters. The outer appearance of the robot "Maria", the "Maschinenmensch" (German for "Machine-Person"), was of great influence for later fictional robots, among them "C-3PO" in "Star Wars". The first industrial robot was a machine called "Unimate", invented around 1956 and used in automobile manufacturing.

Force-Feedback

one form of haptic technology, where an actuators force (usually a motor) is used to convey tactile force (as sensed friction or experienced difficulty) on human input devices (HID). A common application if in racing-simulation steering wheels, but more general, vibration in input controllers may be described as "force feedback" in marketing. Force feedback in robots is an essential technology to make teleoperation of robots and extensions of human limbs (robotic arms) feasible. Force feedback can help with obstacle avoidance (collision), provide feedback on unexpected behavior or mechanics (locked joints). Force feedback is just one form of haptic feedback. Touch sensors are another important aspect of Teleoperation.

Feedback Loop

a control system layout, where a controller's output is feed into an element and at the same time the elements behavior is fed back into the controller as input, to measure and check setpoints of the system. Feedback loops are essential in robots, robotic actuators. A rotational (motor) actuator is usually denoted as a servo motor when it is driven by a feedback loop. More general systems can also be described as feedback loops, for example when a robot evaluates its own actions in real-world space (cmp. Exteroception) and compares actual effects with internal intents and tasks. Compare "Closed Loop System".

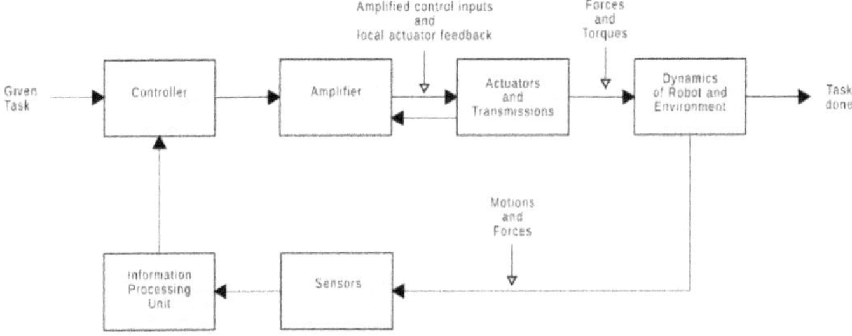

Closed loop robot control block diagram

Gantry robot

one implementation of a Cartesian Robot. A robot which is held (sometimes overhead) by a bridge-like structure that can be moved along two of the Cartesian axes (cmp. "Cartesian robot"), and thus be positioned over a working area for operation. usually, a mounting platform is the central piece of the gantry structure, where either a simple piston like arm realizes movement along the Third (Z) axis, for pick or place operations, or a whole articulated industrial robot is moved by the Gantry, adding translational movement to the Degrees of Freedom of the attached robot.

Gear ratio

In a "gear train" or "gear set", a rotational mechanical linkage, gears of different sizes produce a change in torque, either multiplying or dividing torque. This resulting mechanical advantage or disadvantage is called "gear ratio" and is usually expressed by numbers divided by a colon, for example "2:1" (spoken as "two to one ratio").

Robot Gripper

Like the human arm, a robot usually needs a tool to grasping or holding an object. As the types of tools attached to a robot arm can be manifold, the more generic technical term for a gripper is End Effector. There are different types of grippers:

- Mechanical grippers
- Vacuum grippers
- Magnetic grippers

H-Bridge

is a common type of electronic circuit in robotics. The circuit is able to switch the polarity of a voltage applied to a motor (load) where DC motors need to be run forwards and backwards.

Hand-guidance programming

is an easier approach to robot programming, where the robot operator simply guides a robot or its arm in order to "show" poses and key points of a robots path's and where the end-effector needs to be placed within the robot's work envelope. While "teaching" a robot with a Teach Pendant has come a long way in terms of usability and ease of use, operators or new robot users still fear that a robot needs to be programmed, requiring training and expertise. Hand-guidance programming (usually an optional "mode" in robot control software) lowers the entry barrier for robotics in many applications and helps a more widespread adoption of robots.

Holonomic

The expression holonomic in robotics designates the ability of a robot to make an instant movement in its working environment in any direction without limitations. By definition, a robot is holonomic if its DoF (Degree of Freedom) is equal to the controllable degree of freedom. Good examples of holonomic robots are Castor-wheeled ("Differential wheeled robot") and Omni-wheeled robots.

Homing routine

Kinematic systems may experience positional drift after extended periods of operation. A homing routine is used to calibrate encoders and mechanics to a defined position to reset the (robotic) system.

HRI

short for "Human–robot interaction" (HRI), labels a multidisciplinary field composed of human–computer interaction, artificial intelligence, robotics, natural language processing, design, and psychology that explores the interaction of people (humans, users) and robots. HRI research centers on technical challenges, like motion planning, safety, collision detection but also on psychological or ethical implications of using or interacting with robots. One domain of HRI is the exploration of Roboethics and the Three Laws of Robotics.

Human-Machine Interaction

"Human-Machine interaction" (ambiguously abbreviated as "HMI") describes interaction between a human ("user") and a machine (or computer, or robot), usually via a user interface. While HMI is usually preferred in the context of robotics, a more general or computer-related concept is that of "Human-Computer Interaction" (HCI).

Human-Machine Interface

"Human-Machine Interface" (ambiguously abbreviated as "HMI") describes the "user interface" (UI), a (virtual) space where interactions between humans ("users") and machines (or computers, or robots) occur. HMIs offer (graphical, on screen) displays, gizmos, controls, widgets or (physical, tangible, "tangible user interface") joysticks, knobs, sliders, etc. to effectively and ergonomically control machines, robots or computers, or electronic devices in general. While HMI is usually preferred in the context of robotics, a more general or computer-related concept is that of a "Human-Computer Interface" (HCI).

Humanoid robot

Why build a humanoid robot?

Our modern, artificial world is made for humans. That's why the human form factor is probably the most straightforward way to let robots take over tasks previously delegated to humans, for example in automation (do repetitive, boring or unsafe tasks), without much change in environments. Also, many limitations of current automation result from the bulky, unsafe or different form factor or kinetics of industrial robots.

Why do humanoid robots have legs?

When nature invented the leg, it produced the most versatile tool to navigate our world. A leg will, simply put, allow you to reach and go more places than a wheel or a caterpillar tread, while a being more energy efficient than wings, spinning rotors, which are also unstable and fragile. A leg can take steps, it can climb, stairs or ladders, it can be used as an optional arm replacement, etc.

Humanoid robots in 2024

A number of companies are working on humanoid robots in 2024. Combining cutting-edge artificial intelligence, reverse kinetics algorithms and self learning systems leads to impressive results. Here is an incomplete list of interesting humanoids in the works:

- **ASIMO** developed by Honda
  ASIMO being short for "Advanced Step in Innovative Mobility", was first presented in 2000 and received a number of updates over the following decade.
- **Nao** developed by Aldebaran Robotics / SoftBank Robotics
  is a roughly kid-sized programmable humanoid robot popular in academia, research and education. It is well know for replacing Sony's dog-shaped Aibo robot in the RoboCup [en.wikipedia.org/wiki/RoboCup] soccer challenge.
- **Atlas** from Boston Dynamics
  is a variable sized bipedal humanoid robot currently un-

der development with impressive dexterity. Various
U.S. research and military institutions are involved in
its development.

- **Optimus** in development by Tesla
  The "Tesla Bot" now nicknamed 'Optimus' (not Optimus Prime from Transformers) is a general-purpose robotic humanoid, designed for assistance, domestic and automation work.

- **Leonardo's mechanical knight** by Leonardo da Vinci is probably one of the earliest incarnations of a man made humanoid robot. It is an automaton, not a real robot, a number of wheel, cables and pulleys that move a medieval armor, giving the impression of a person inside moving arms and upper body.

Hypoid gearbox

is a special subtype of "spiral bevel gear" (or more general "bevel gears") where the pinion is offsetted from the axis-plane of the crown (axes are non-intersecting and not parallel). The offsetting of the pinion has the effect that the gear train is able to transmit higher torque in comparison to a (spiral) bevel gear assembly. Teeth on the gears are usually helical in more sophisticated designs, resulting in less noise and less wear. Hypoid gearboxes are commonly used to rotate (industrial) robot joints. A simpler, related gear train is a Worm gear drive.

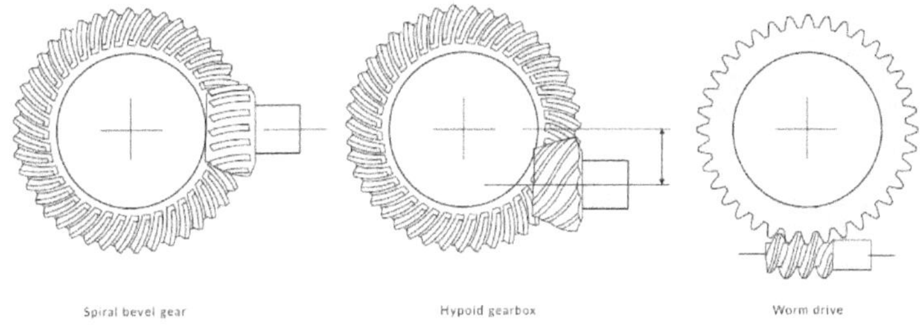

Spiral bevel gears, a hypoid gear train and worm gear drive

Industrial Robot

The term "Industrial Robot" is usually used to describe robots used for (industrial) manufacturing. Such robots are optimized for high speed, repeatability and minimum wear, are sealed against the ingress of liquid or dust and designed so that their work envelope is maximized or tailored to a specific need.

Industrial Robot Manufacturers, by Country

- China: KUKA
- Japan: Fanuc, Denso, Epson, Kawasaki, Mitsubishi, Yaskawa Denki (Motoman), Omron
- Switzerland: ABB, Stäubli
- USA: Adept Omron

IMU

an IMU (inertial measurement unit) is an integrated type of sensor, usually a combination of an accelerometer, gyroscope and sometimes a barometer and magnetometer. IMUs are used to

39

measure a body's specific force, angular rate and sometimes its orientation in (Cartesian) space. IMUs usually employ some form of on-chip or on-board "Sensor fusion", significantly contributing to their accuracy and reliability. Compare "sensor fusion". IMU play an essential role in land-based, marine, aerial and space navigation, of aircraft, missiles, UAVs, vehicles and robots. Capabilities of IMUs are given in Degrees of Freedom (DoF) and performance of the different sensors, error rate, resolution, frequency, etc.

Industry 4.0

also "Fourth Industrial Revolution" or "4IR", is a (marketing) term to describe initiatives and projects on the way to a complete digitization of industrial manufacturing and related (business) processes. It originated in Europe, Germany, and in circles of the World Economic Forum around 2015. As of today, only a few years later, it already became common to speak of "Industry 5.0", taking the "human factor" of manufacturing back into account, as the original outline had a too tech centric perspective on things.

IP code

Robotic IP ratings refer to the IP code for equipment, with "IP" being short for "ingress protection", a code to describe how well a device is sealing in or protected against external water and/or dust. For example, clean-room robots must be dust tight and to a certain degree water proof to prevent any contamination of the operating space by the robot. IP rating **IP67** is common for clean room robots, but also for foundry robots, where protection against dust from the environment is more important than in clean-rooms. Other robots are able not only to immerse into water for some time but to perform under water extensively. Many industrial robots have different IP ratings for their body and their arm, meaning that, for example the body is IP54 rated while the arm is IP76 rated, as protection classes are difficult and costly to achieve and a certain application might expose the robot to critical environments on in parts. It is common that robot vendors offer to upgrade robots in IP rating via optional product packages.

Example "ingress protection" (IP) codes:

- **IP00**

  This code means that there is no protection at all. A typical example is a dry-type transformer without any enclosure. The transformer would be installed indoors in a bay having a door in order to prevent accidental contact with live parts

- **IP53**

  This is the minimum IP code that should be used for equipment installed outdoors. IP54 would be a better choice since the second numeral 4 means protection against water coming from all directions whereas the numeral 3 means protection against water coming from an angle of 60°

- **IP67**

  The first digit "6" means a robot is dust tight and completely protected against the intrusion of dust. The second digit "7" means the robot can be immersed into water at 1 meter pressure level for up to 30 minutes (temporary water immersion). IP 68 would mean the robot can withstand the effects of continuous immersion in water (without the actual pressure level remaining undefined).

What are Kinematics?

Kinematic calculations or kinematic equations are used to mathematically determine (compute) a robot's end effectors position, when some or all of a robot's joints are moved, or, reversly, to which positions joins have to move in order to position the end effector at a set coordinate. Kinematic calculations to deduce the position and orientation of a kinematic system, a kinematic chain, from a number of joint angles are called forward kinematics. For example to calculate the end-position of a robot arm or its end effector from the angle values of its joints. Inverse kinematics, in turn, is the process of calculating the joint angles needed to place the end of a kinematic system, a kinematic chain, to a set coordinate position and orientation. For example, inverse kinematics are used to calculate the needed servo motor positions for a robot arm, so that its end-effector reaches a given coordinate in a given pose, relative to its base, with the base being the start or root of the kinematic chain. Inverse kinematics are also frequently used in 3D animation, camera tracking for film and video and in game development.

LiDAR

short for "Light detection and ranging", or "Light imaging, detection and ranging", is a method similar to RADAR for optical remote sensing, of distances and speed. It is a form of 3D laser-scanning. Instead of radio-waves as with RADAR, LiDAR uses laser light and time-of-flight measurements. Contrary to optical vision as in cameras, LiDAR is an active method of scanning the environment. As such, and depending on used laser wavelength, it may interfere with IR receivers or be harmful to organisms. Commonly deployed LiDAR systems use low energy lasers from a spectrum not visible to humans.

Link

A robot is made up of links and joints. A link is the (usually) inflexible structure between the joins of a robot. If we think of an articulated robotic arm, the arm is made up of "bones", the "links", and the flexible parts, the "wrists" or "joints", that allow the links of the arm to rotate and move.

Limit switch

an electro-mechanical limit switch is a physical (micro) switch that is used to restrict the movement of a robot (or machine). It triggers a signal or breaks a power circuit for a motion to immediately stop once an actuator has driven some form of movable element to an end, to an end-stop position or beyond. Limit switches are usually used to maximize the available reach of a drive system while at the same time preventing damage to the robot itself, surrounding infrastructure or to prevent injuries to humans around. Often, robotic drive systems differentiate between a "soft stop" and a "hard stop". A soft stop is usually software implemented and represents a limit the system tries to avoid during normal operation. A hard stop may trigger once the system behaves outside normal operation, for example when outside forces move a robot's arm beyond a defined limit that's normally not reachable by the robot. Hard stops are usually designed in such a way that once fired the robot can't easily resume operation or be easily switched on again, as reaching the hard stop signaled a serious malfunction. In some robots, a physical cap or breaker gets bursted off its holder and restarting the robot requires intervention by a technician. Optional limit switches, added to a robot in addition to the built-in limit switches, may be used to restrict a robot's work envelope and form a subset volume of the work envelope, the "operating envelope".

Robot Locomotion

Just like any mobile thing, a robot, if not designed as a stationary system, may be mobile to move through its environment. Commonly, a robot moves on wheels, tracks, legs or wings, having the option to roll, glide, walk or fly. Via bionic imitation, a robot may use spheres as a variant of wheels, use climbers, suction cups or similar to climb on trees or walls, may move in a snake-like rocking motion, use flippers or fins in water and use rotors or fixed-wings to fly.

Linear Actuator

is any mechanic that creates a linear motion (straight motion stroke). Linear actuators may be driven by an electric motor, hydraulics (hydraulic cylinders), pneumatics (pneumatic cylinders) or piezoelectric. Implementations using a motor usually convert a rotary motion into a linear by means of a gear or mechanic linkage. Different implementations exhibit different traits in terms of accuracy, wear and energy to torque efficiency.

Comparison of linear actuator mechanisms

	Leadscrew	Rack and pinion	Belt drive	Linear motor
Positional accuracy:	average	low	low	very high
linkage efficiency (energy to torque):	very high	good	low	high
Possible length:	short	medium	short	long
Speed	average	high	high	very high

Linear drives in storage devices

Rotating media storage devices essentially work like a record-player, where a pick-up arm (stylus) is moved over the record spun by a turn-table. With magnetic media, as in floppy disk drives and hard disk drives, there are no grooves to follow (in FDDs and HDDs) and, in addition, the head usually doesn't touch the surface (HDDs). That's why in such devices, the read/write head is positioned over tracks via linear drivers. Early floppy disk drives, like the 5.25" Micropolis 1015 [www.micropolis.com/1015.html], use a screw drive, with a leadscrew and stepper motor, to align the head with tracks on media. Smaller 3.5" floppy disk drives usually employ a variant of the belt drive, resembling a rigid chain actuator, where a bent threaded strip of metal is moved by a motor, like in a rack and pinion assembly, but moving the metal-strip around a roller to move the read-write head sled. Winchester-style hard-disk drives, e.g. Micropolis 1320-series 5.25", Micropolis 2217-Series, usually use Voice-coil driven, Galvanometer-like mechanisms to position r/w-head arms in a more circular movement. Galvanometer/voice-coil actuators move iron by a magnetic field established by an electric current flowing through a coil - changing the applied current moves the iron core accordingly.

Linear Motor

While rotary motors have a circular rotor (armature) core turning inside a concentric stator, linear motors have this structure flattened out (unrolled) on a linear track or along a tube. A linear motor is sometimes called "linear synchronous motor" (LSM) or linear servo motor. There are many different designs and implementations among linear motors, with differences in exhibited traits. usually, linear motors can be distinguished in 1) low-acceleration, high speed and high power motors, "linear synchronous motors" (LSM) and 2) high-acceleration linear motors, "linear induction motor" (LIM). Some circular motor designs used in "direct drive" mechanisms (e.g. as used in turntables and washing machines) are also called linear motors due to their rotor/stator layout. Some linear motors are not only able to exert a linear force, but can also turn their core - these combination motors are called "linear rotary motors".

Caterpillar tread

A caterpillar tread, "tank tread" or more formal "continuous track", "chains", chain track or tracked treads are a system of vehicle propulsion used in tracked vehicles like heavy machinery, tractors or military vehicles. A tracked thread surpasses traction on muddy or slippery surfaces ("all terrain") and has a number of benefits like improved weight distribution, but is often heavy, stiff and difficult to maintain. One example of a robot having a caterpillar tread is "Johnny 5" from the 1986 U.S. feature film "Short Circuit". Machine vision Machine vision in robotics refers to a technology that provides visual perception for a robot's decision-making. It enables robots, particularly autonomous vehicles (AVs), autonomous mobile robots (AMRs), and drones, to recognize obstacles and find suitable paths in their working environment. Visual data is collected by cameras and special sensors and processed using image processing algorithms. The resulting data is then processes by the robot's controller and generates commands for the robot's actuators, like motors and servos, to execute specific tasks. In today's robotics (as of 2024) the most common uses of machine vision in automation and robotics are visual inspections and defect detections, positioning and measuring parts, pick-and-place operations, three-dimensional mapping, etc. Future robotics will expand machine vision capabilities, as vision is such an important sense for humans, animals and any lifeform on earth, and robots will probably align with these capabilities.

Mecha

sometimes simply a "mech", is a term used in science-fiction, videogames, anime and mange, to describe fictional tall, often armoured and armed, giant bi-ped walking robots or machines, usually piloted via neuro-interfaces or intricate mechanical harnesses. The word "Mecha" is a Japanese neologism, from shortening Japanese "mekanizumu" (English for "mechanism"). Popular fictional Mechas include the assault machines from BattleTech, ED-209 from RoboCop, the AT-ST (all-terrain scout transport) Walker from Star Wars and the Amplified Mobility Platform (AMP suit) from Avatar, with similar hydraulic mechanized walkers in the Matrix movie and Halo videogame franchises.

Mechatronics

is an interdisciplinary engineering branch that combines (broadly put) the disciplines of mechanics/mechanical engineering, electronics/electrical engineering and computer science/information technology. These intersections link the discipline to even more sub-disciplines, such as Smart Structures ("adaptive electronics", "Adaptronik"), electromechanics, precision engineering, microsystems technology, optoelectronics and optomechanics. Mechatronics are an integral part of any robot design.

Mobile robot

describes any robot that is able to move around (is not stationary), either in a specific environment or the open world. Mobile robots are most commonly ground-based, but can just as well be robots that are able to fly or dive. Robots navigating the ground are usually a little less complex to control, as a swimming or flying robot has to be able to control one more spatial dimension ("3D trajectory planning"). To be mobile, this type of robot has to employ some form of locomotion mechanism, ranging from wheels to articulated legs. Mobile robots need to be aware of their environment, be able to plan their movement in 2D or 3D space, have some form of collision detection implemented ("object avoidance") and need to make sure that their movement doesn't have safety implications for people or things around them.

Motion Controller

Usually refers to a sophisticated form of DC Motor Controller. The aspect of "motion" here refers to the fact, that a motion controller doesn't merely drive current to a motor to spin it at arbitrary speeds, but is optimized to quickly and precisely adapt current to sped up or slow down the motor, apply a form of soft brake, control spin up current, control direction reversals, etc.

Motion System

Usually a ready-made assembly of motors, actuators, mechanics, encoders and controllers that is offered as an off-the-shelf product solution by a vendor.

MTBF

short for "mean time between failures". It's a measure that describes the average duration of normal operation of equipment in-between two failures.

MTTF

short for "mean time to failure". MTTF is a statistical measure that describes the average duration of normal operation of equipment until failure; IEC 60050 (191)'s exact wording is: "the expectation of the time to failure". It is no guaranteed lifetime assessment and more of a guiding value gathered from observation and testing. There's also a related value, $MTTF_D$ (MTTFD) to describe the "mean time to dangerous failure".

Newton

Netwon is a unit to define force. One newton is the force needed to accelerate one kilogram of mass at the rate of one meter per second squared (meaning increasingly "accelerate at a rate of one meter per second per second"). The symbol for newtons is a capital "N". Compare "Torque", which is measured in (metric) Nm (newton meters).

OCU

"OCU" is short for "Operator Control Unit", and is similar to a "teach pendant", as it is a (usually) hand-held input device used to control a robot, drone or mobile platform. OCUs are tethered to a robot or its control unit through a cable cord, fibre-optic link or tx/rx radio connection. Depending on actual physical implementation, the OCU might be part of a software suite and be run on a computer, like a laptop, which would then be named a "laptop control unit" (LCU) or in a military context a "tactical robotic controller" (TRC). All devices usually share a rugged physical design, comparable lightweight, to make the device simple and convenient to use, as is important in search & rescue, medical or military applications.

Omnibot

family of popular 1980s amateur toy-robots produced by childrens toy manufacturer TOMY.

Operating Envelope

sub-volume of the work envelope of a robot. While the physical work envelope of a robot is defined by its size, degrees of freedom and rotational or linear capabilities of its axes, the "operating envelope" is a sub-volume of the work envelope enforced virtually through software defined limits and/or real-world limit switches. Compare "Reach".

Patrol robot

Patrol robots are one application of robotic technology where a mobile platform (mostly ground based, less often airborne) is equipped with sensor arrays suitable for surveillance, recognition and detection of objects and people in a defined area. The idea is to deploy systems that are able to fulfill duties normally performed by human law and security enforcement personnel, of private security firms, warranted law employees like police officers, private military companies (PMC) or other government or non-government organization personnel. As of 2024, stationary robotic PTZ surveillance cameras in combination with intrusion detection (motion detectors, light curtains, nighvision devices, etc.) mechanisms are much more common than **Patrol robots**. In future applications, advanced robotics are expected to (in part) be able to prevent, detect and report crime and be a general assistance to the general public by helping maintain the public order. While even supporting tasks are currently (as of 2024) mostly planned, it can't be said for sure when or if at all patrol or security robots will be able to replace human officers in law enforcement, as this is one of the most challenging job roles even for human cops. For example, the apprehension of suspects, discerning lawful from unlawful actions, morale and ethic arbitration (machine ethics) is at least very difficult for current artificial intelligence systems. As such, a thing like armed patrol robots is as of now (2024) quite unthinkable. Despite that, some states and large cities are researching or boldly implementing robots in their future plans. Robotics in Dubai e.g. was an emerging topic. In 2017, a humanoid patrol bot was presented in the United Arab Emirates, with the ambitious aim to patrol cities and make communities safer through AI. The presented police officer robot was a cus-

tomized model REEM-C from Pal Robotics Dubai the area of
its duty.

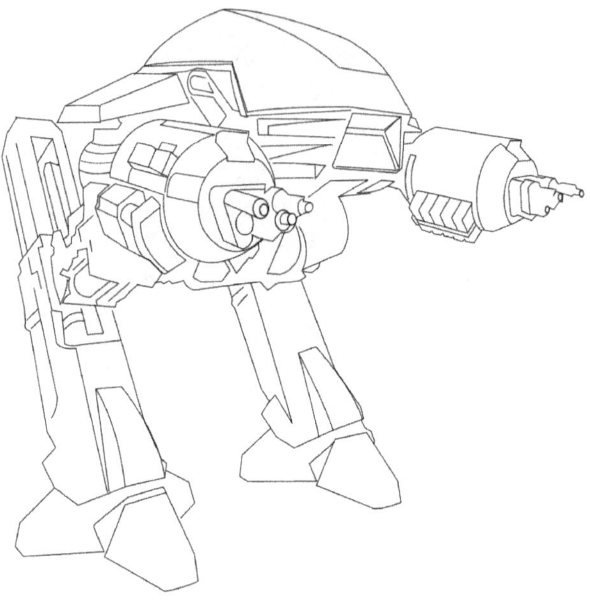

*A bi-ped **Patrol robot**, as imagined in 'RoboCop', a dystopian science-fiction feature film. Armed and out of control. See 'Robots in popular culture' and 'Mecha'.*

Payload

the amount of weight a robot (arm) can handle or carry while staying inside the system's defined tolerance, e.g. for accuracy, repeatability, MTBF, etc. Payload is differentiated between "Maximum Payload" and "Nominal Payload". Maximum payload defines the amount of weight a robot can handle at reduced speeds so that the effect of gravitational and inertial forces remains at a minimum. Nominal payload is the amount of weight a robot can handle at its maximum speed. Due to the various shapes of handled objects, form drag of large but light objects moved at high velocity may influence a robots payload.

Perceptual system

mechanism designed to provide a robot with the ability to sense, understand, and interpret its external and internal environment. The main components of such a system are various sensors and sensory data, machine learning algorithms and data representation in the form of environment modeling. The Perceptual system allows robots to make informed decisions and interact accordingly. It includes proprioceptive and exteroceptive sensors that provide necessary data about the robot's internal state and the surrounding environment.

Planetary gear

a "planetary gear", more precisely, a "planetary gearset" or "epicyclic gear train" is a type of gear assembly where a central toothed shaft ("the sun") is orbited by two or more gears ("planets") and held together by a third outer toothed element, the outer ring gear. Planetary gear trains are used as gear reduction or planetary reduction drives, to shift the rotational speed between two shafts. The planetary gear was invented by the Scottish engineer William Murdoch around 1800.

PLC

short for "Programmable Logic Controller". PLCs are programmable controllers, a type of industrial computer or microcontroller that has been ruggedized and adapted for the control of automation equipment, actuators and robots. Dedicated operating systems, like "STEP 7" (compare "S7") enable the controller to act as a "hard real-time" system, where inputs are processed and answered in a defined, very short interval. In Europe, the Siemens SIMATIC SPS is a very popular system of PLCs, with other known vendors being Allen-Bradley/Rockwell Automation, Schneider Electric, Omron, Mitsubishi, Bosch Opcon, Bosch Nexeed. PLCs are commonly mounted on DIN rail [www.micropolis.com/support/kb/rack-mounting-faq#DIN-Rail] metal mounting rails.

Polar Robot

also called a "Spherical robot", the Polar robot is defined by a horizontal arm that is linearly extractable and resting on a rotatable and tiltable base. This way, the robot is able to reach back and forth and move objects or the end effector to lower or higher coordinates within its spherical work envelope. Some of the earliest industrial robots were polar robots. The Unimate is the most famous example.

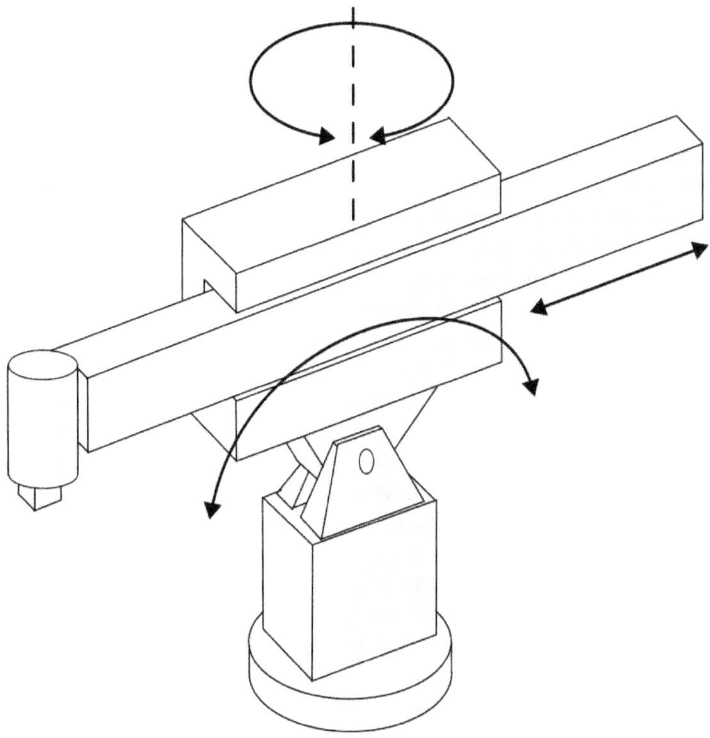

Polar Robot configuration illustration

Pose

"pose" is the combination of position and orientation. A robot can have or drive to a "pose", a robot wrist can have a pose and a robot's end-effector can have a pose.

Programmable logic controller

see "PLC".

Proprioception

Proprioception (from Latin 'proprius', "own, dear") is a robot's ability to sense its own internal state by processing data collected from its various sensors, like accelerometers, gyroscopes, motor position sensors, joint encoders, etc. Unlike proprioception, exteroception means a robot's ability to sense its external surroundings. Together, proprioceptive and exteroceptive sensors form the perceptual system of a robot.

PTZ camera

A "pan tilt zoom camera" (PTZ Camera) is a robotic appliance commonly used in remote video systems, like **surveillance cameras** or for TV broadcasts where room is either limited, making a human operator difficult to place, or where a local human operator is unwanted. Operating a PTZ camera is done via a remote control desk (Teleoperation), its name is derived from the channels or axes the camera is able to control remotely, the pan and tilt axes and the zoom actuators. Most PTZ camera actually allow to remotely control much more parameters than these three.

PWM (pulse width modulation)

method for controlling the power delivered by an electrical signal, effectively managing the quantity of total power delivered to electrical components, such as motors and actuators. While a constant signal amplitude delivers constant power, in PWM the signal is switched on and off for defined periods. The longer the switch is on, the more power reaches the attached device. This method of PWM finds applications in many areas, for example in motor speed control or LED brightness adjustment. In robotics, the term "PWM" is often associated with actuators based on hobbyist RC servos. Here, the pulsed signal isn't used as such, but to encode values via the pulses' duration. The analog resistor value of a potentiometer is encoded (modulated) into a signal transmitted by a RC radio, then again demodulated (decoded) in the receiving servo's controller to receive a setpoint for the servo's position. More on RC Servo Control in Wikipedia [en.wikipedia.org/wiki/Servo_control].

Reach

usually expressed as "Reach Envelope" or "Work Envelope", is a complex metric to measure and display the maximum distance a robot (usually a stationary industrial robot) can reach from its pedestal.

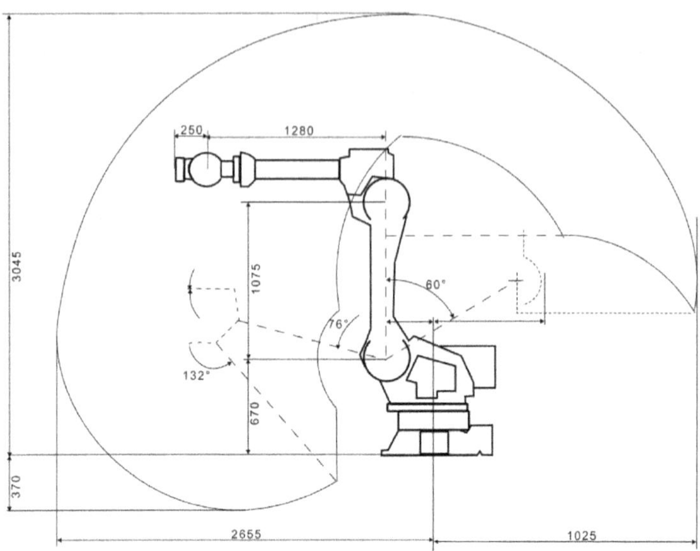

Diagrams like this one usually illustrate a robot's specified work envelope.

Repeatability

describes a robot's ability to repeat the same motion (reaching certain setpoints) when driving the robot with the same control commands or signals as in a previous iteration ("cycle-to-cycle error") and with the same payload on the end effector. Repeatability is defined by manufacturers for a robot within its work envelope and for payloads within its payload capacity. Repeatability can be rephrased as the expression of Accuracy in a repeated motion (cmp. Accuracy).

58

Resolution

defines the granularity of sensors, feedback-loops and actuators, and measures the smallest increment of motion that a robot can detect as part of its proprioception (sensor data from encoders) and the smallest motion the control system is able to drive.

Roboethics

Roboethics, as an abbreviation of "Robot ethics" describes a branch in general ethics concerning with the creation and use of robots and robotic beings, as well as with questions of how humans interact with robots, what robots should be allowed to do or are expected to do, what their role in everyday life is and should be. Thre is a long history of discussion of robot ethics, in academia as well as in fiction. Outcome of such discussion are things like Asimov's Three Laws of Robots or legislation regarding the creation of robot soldiers or the armament of robotic platforms, regardless if the are teleoperated, partly autonomous or fully autonomous (Lethal Autonomous Weapon Systems, LAWS).

What about Robot Soldiers?

Today, most robot researchers and developers take a very strong stance against the weaponization of robots in general or the armament of robots specifically. World-wide, legislation is currently developed or put in place to ban armed robots or prevent robots from doing harm, from exerting physical force against humans to being able to severely hurt people. In current politics, there are weapon systems that are banned and states seem to obey set rules on some systems, while other areas receive less compliance. Armed airborne drones are one controversial area in politics. Regarding robots, as sad as it is, it it's probably safe to assume that no technology will only ever be used for peaceful things alone.

Robots in Popular Culture

Most famous movie robots

Of course, any list compiling the 'most', 'top' or 'best' robots in movie history can never be complete or free of bias. Which robots rank among the most popular is hard to determine, or which robot is 'the best' is equally hard to measure as it is deeply subjective. That said, here's a list of some well-known robots of film history:

- **R2-D2 and C-3PO** from Star Wars
  The inseparable robot couple, the Laurel & Hardy of Space Opera, the small handy droid and the awkward golden translator robot from the Star Wars franchise.

- **Johnny 5**
  In Short Circuit (1986), lightning strikes and sparks human-like intelligence in an armed military robot. The robot escapes and befriends with a group of humans. His overall appearance, designed by industrial designer Syd Mead, and his caterpillar treads clearly were the blue-print for Disney/Pixar's cute **Wall-e**.

- **Ava** from "Ex Machina" (2015)
  represents the contemporary robots and the Gynoids on this list - and Ava is very human-like, measuring the uncanny valley [en.wikipedia.org/wiki/Uncanny_valley] to unknown depths.

- **Data** from the Star Trek: Next Generation
  The idea humanoid, polite and humble, eager to learn and help. Lesser know fact: his positronic brain [en.wikipedia.org/wiki/Positronic_brain] was dreamed up by Science Fiction author Isaac Asimov, not Gene Roddenberry.

- **ED-209** from RoboCop
  a tall bi-ped mecha, the *enforcement Droid series 209 is a self-sufficient law enforcement robot, programmed for urban pacification*, is what fictional marketing tells the audience in this dystopian feature film. After that it goes downhill very quickly and the audience learns what harm armed robots can do. See **Patrol Robot**.

- **Bishop** from "Aliens" (1986)
  unlike 'Ash' from the Alien prequel, the James Cameron version of a humanoid robot is kind and humble while surpassing any human in dexterity and sense of comprehension. He the Data of the Alien universe.

- **T-800** from "Terminator" (1984)
  While mankind actually explores what's possible with artificial intelligent, the arts, films, have to explore what the future might bring, ask the essential questions. Jim Cameron invented the Terminator, a mean determined utility made for war, only to re-invent the Terminator as a humorous, shapeable machine ready to support us humans.

Robot

A robot is a machine designed to replicate certain human movements and execute specific tasks in the real world. Industrial robots work like automatons, not "knowing" about their surroundings. More elaborate robots are able to sense their environment and feed computations with this input, allowing for limited autonomous decision-making, without direct human control.

Robot Operating System (ROS or ros)

not an actual computer operating system but rather an application framework and set of software tools and libraries providing the functionality of an operating system. The framework helps develop, test and deploy robotic software.

Robotic Exoskeleton

is a typo of robotic prosthesis for humans that is applied to the exterior of a human body, to support the body's own muscles and bones and/or protect the wearer from injury or stress. Different types of Robotic Exoskeletons have been devised. Some exoskeletons are meant to give disabled persons abilities back they lost in an accident or have lacked from birth. Other exoskeletons are meant to amplify the human strength or endurance, in challenging or dangerous work environments, for example. Some Robotic Exoskeletons are delicate mechatronic devices modeled after the contours of the human body, while others are designed as a large tank-like structure meant to withstand large forces. Mechas and walkers in fiction are examples of the latter.

Robotics

Robotics is a branch of engineering and a multidisciplinary field concerned with the study of robots. It includes mechanical and electrical engineering, computer science and artificial intelligence. The field focuses on a robot's (autonomous) sensing and acting in the physical world. The purpose of robotics is the creation of smart machines, intended to assist humans in numerous ways, like automating repetitive and monotonous tasks or performing actions in dangerous or challenging environments.

RS232

formally written with a hyphen as "RS-232" ("Recommended Standard 232"), is a standard originally introduced in 1960 for serial short-range, point-to-point, low-speed wired data connections communication. It was developed by the "Electronic Industries Association" (EIA), the same standardization body responsible for working out the 19" rack specifications. The official current standard is revision "F" (as of 2024) and is named "(ANSI EIA/) TIA-232-F", dated October 1997 with amendments until 2012. In practice, computers stopped being shipped with a serial COM port since around 2010 and the protocol standard started to disappear in favor of more robust and faster connection buses (like USB). An exception being industrial PCs, POS terminals and industrial applications, where RS-232 and its common DB9 connector is still in widespread use, especially where long cable lengths and tolerance against interference is of importance - a robustness stemming from RS-232's comparably high signal level. Connection-wise the full RS-232 specs define 9 wires, with elaborate Handshaking and Control signaling, yet in reality, these have traditional irrationalities designed in, are thus often miswired in implementations and resulted in being seldomly used. Today, most devices communicating via RS-232 only use three wires: RX, TX and Ground/

GND. One popular field of application for RS-232 in recent years has been the maker movement, with Arduino and other microcontroller platforms using RS-232 serial connections for programming and communication.

Related reading:
- Is RS-232 the same as RS-232 C? [www.micropolis.com/support/kb/micropolis-bbs-faq#RS232-vs-RS232C] from the Micropolis BBS FAQ
- Rack Mounting FAQ: Console-Server [www.micropolis.com/support/kb/rack-mounting-faq#Console-Server]

RS485

formally RS-485, is a standard originally introduced in 1983 for serial communications systems. It defines the electrical characteristics of drivers and receivers for use in RS485 and added multi-drop bus and multi-point capabilities to older standards, most prominently in comparison with RS-232.

RTLinux

RTLinux was an alternative Linux microkernel converting Linux into a "hard realtime real-time operating system" (RTOS), with the intent of enabling Linux to be used as a control OS. It was later acquired by Wind River Systems and then discontinued in 2011.

S7 (STEP 7)

STEP 7, commonly abbreviated as "S7" is a programming language for Siemens' SIMATIC-S7-family of Programmable Logic Controllers (PLC).

Self-managed

a term emerging in the field of robot control software theory. Real world observation of robot software has shown that many situations can't easily be handled by traditional rigid software designs. One solution might be adaptive software that reconfigures ("self manages") itself during runtime, rearranges specific modules of its subcomponents or adapts program flow to find answers to unforeseen problems.

Sensors

a sensor (from Latin "sentire", meaning feel or sense) allows a robot to determine its current internal and foremost external state in its environment. Choosing the right sensor is one important aspect in the design of robots. Sensors vary in terms of size, technical complexity or complexity in their support (power, stability), data precision and cost. Types of sensors are usually infrared (IR) sensors, sonar, laser and optical (camera) sensors.

Sensor fusion

is the process or computationally merging the output of multiple sensors into one value or set of values. Often, sensor fusion is used to improve accuracy by smoothing sensor output and/or eliminating noise. Sometimes sensor fusion is used to form a different, more complex kind of sensor data from a number of sensors that are only able to produce a simpler form of sensor data on its own, for example, vector data from an IMU (inertial measurement unit) which in turn is again a combination of different sensors, an accelerometer, gyroscope and sometimes barometer and magnetometer.

Singling station

In automation, many intermediates or blanks come in unsorted bundles, in chaotic orientation on conveyor belts or in bulk, from boxes or bags. Traditional industrial robots only have limited capabilities to adapt to positional change of handled objects. In automation, robots rely on the environment within their work envelope to remain unchanged so their programmed movements lead to predicted results. A singling station sorts bulk material and arranges single individualized objects in a predictable manner for further processing (cmp. "Centering Station").

Singularity

Robot singularities are limits imposed on the robot and its reachable work envelope or its possible poses by physical limits of its wrist / axis movement. It's a problem common with multi-axis articulated robot arms. When arriving at a singularity, due to how wrists are physically constructed, the robot loses its ability to move around one or more axes. One can think of a singularity as a point on a virtual plane that constitutes an extreme pose for one wrist, close to a physical limit, and although the robot is physically able to reach points behind this plane, it usually requires driving some or all other joints to a different pose so that points behind the plane can be reached again. As such, "going through" a singularity is a complicated maneuver for a robot arm. Singularities usually occur, when an articulated robot arm with rotational wrists has to move its end-effector along a Cartesian coordinate system path. "Dead zones" are also singularities. "Gimbal lock" is one type of singularity found in (camera) gimbals.

> Related material:
> - Robot singularities demonstration
>   [www.youtube.com/watch?v=ID2HQcxeNoA] (YouTube Video by Mecademic)

SLAM

short for "Simultaneous localization and mapping", a navigation technique used by robots and autonomous vehicles. Different sensing technologies are used to map an environment (laser ranging/ LiDAR, 3D sonar sensors or 2D cameras) while at the same time calculating the position of a robot or vehicle within this mapped space.

Sonar

the term "Sonar" is an acronym of "Sound navigation and ranging" and describes a type of sensor that uses echoes (waves) in a water medium and runtime measurements to determine the distance of obstacles and surfaces. Building on the well-known marine sensor, it is common to speak of sonar for ultrasound ranging in land-based robotic applications.

SCARA Robot

"SCARA" is short for "Selective Compliance Assembly Robot Arm" or "Selective Compliance Articulated Robot Arm" and describes a special type of robot that is optimized for assembly and pick and place operations. Its design is centered on horizontal transportation tasks with supplemental vertical place actions. A SCARA robot resembles a human arm as it has a two rotational joints layout, thus the "articulate" in its designation. Some SCARA robots use two linked arms working in tandem to improve payload capacity or stability. Both types allow the SCARA to move the end effector on the X and Y axis within the work envelope, but joints are fixed on the Z axis. This fixed Z axis allows the arm to be precise in XY and/or carry larger weight. This rigidity on one axis led to the name part "Selective Compliance". Usually, the third Z-axis is implemented as a stamp like linear action tool, allowing the robot to lift or place objects.

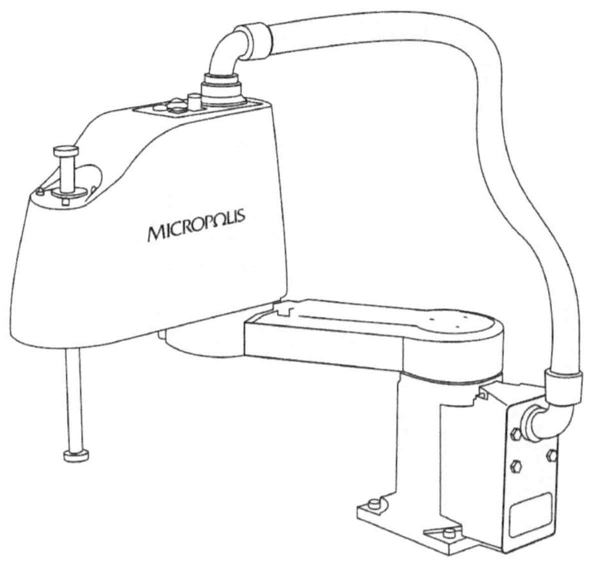

SCARA robot as commonly used in pick and place automation

Servo Motor

Combination of a motor and encoder, forming a closed-loop system, able to control position and speed of the drive system. A servo motor, embedded in a more elaborate system with mechanics and dedicated controller electronics can be described as a "motion system".

Standard gravity

Established by the 3rd General Conference on Weights and Measures in 1901, the "standard gravity" (also more verbose "standard acceleration of gravity" or "standard acceleration of free fall") is the nominal gravitational acceleration of an object in a vacuum near the surface of the Earth and is used to define the standard weight of an object (as the product of its mass and this nominal acceleration called "standard gravity"). "Standard gravity" is denoted by g_0 or g_n, a constant defined by standard as 9.80665 m/s^2. This value, for example, is used in equations to calculate motor torque required to hold or lift a weight. 1kg * 9.8 = 9.8N. Compare "What is the torque required to lift a certain weight?".

Stepper Motor

a special type of electrical motor that doesn't turn its shaft continuously, like brushed DC motors do, but instead rotates its axis in small incremental steps. Internally, this is facilitated by arranging internal magnets in a "toothed" pattern, where the similarly "toothed" rotor only increments from one tooth to the next instead of doing full turns. Stepper motors (also "step motor" or "stepping motor") are a common type of robotic actuators.

SPA architecture

The SPA architecture, short for "Sense Plan Act", is a robot behavior paradigm for robots. Getting popular during the 1960s, the idea is to first sense an environment, then planning a action within it, using an abstraction of the environment, and finally, acting based on the set plan. Sometimes, this is described as the "Sense, Plan, Act"-Cycle or "SPA-Cycle", as this is repeated over and over. Since about the 1980s, this paradigm has been marginalized in favor of refined models, usually trying to reduce the computational distance between current state and goal.

In some models, this is achieved by breaking the SPA cycle into active modules, components that communicate with each other but solve different aspects of a robot's movement in parallel.

Teach Pendant

A teach pendant is a portable robot control device designed to provide an interface a human operator can use to program applications, control motions and teach robots, usually while standing near-by and guiding the robots actions.

Teleoperation

Teleoperation in robotics refers to the technical term that designates remote control of robot movements by a human operator, utilizing hand controllers, joysticks or haptic devices. The use of various technologies in teleoperation provides direct human intervention in impractical or dangerous situations without the physical presence of human operators.

Three Laws of Robotics

is a (fictional) set of rules invented by science-fiction author Isaac Asimov that were central guidelines for robots in a number of his works, the first work being the short story "Runaround" (1942) where these rules are presented as being circulated through the "Handbook of Robotics, 56th Edition, 2058 A.D.". Since their inception, the three laws were adopted and altered in other authors' works of fiction as well as in real-world academia in relation with musings about robot ethics and ethics of artificial intelligence (machine ethics). Though originally invented as being rules for robots, Asimov later regarded the three rules as being so universal that they also should apply to human behavior. The original three rules are:

- The First Law: A robot may not injure a human being or, through inaction, allow a human being to come to harm.
- The Second Law: A robot must obey the orders given it by human beings except where such orders would conflict with the First Law.
- The Third Law: A robot must protect its own existence as long as such protection does not conflict with the First or Second Law.

Related reading:
- Three Laws of Robotics
  [en.wikipedia.org/wiki/Three_Laws_of_Robotics] in Wikipedia
- Laws of robotics [en.wikipedia.org/wiki/Laws_of_robotics] in Wikipedia

Timing Belt

a "timing belt" ("cogged belt", "cog belt", "synchronous belt", "Gilmer belt" or sometimes "cambelt") is a toothed rubber belt, used to transfer drive from one (belt) pulley to another. Toothed belts are common in robot arm drive mechanics / transmission systems. Rubber belts in general are perishable components, yet bear a number of advantages, like low cost, low friction loss, low noise and belts do not require lubrication. The use of toothed belts and cog wheels has advantages over plain rubber belts and smooth pulleys, as the teeth prevent the belt from slipping and thus increase mechanical linkage. Although invented for use in textile mills around 1940, the name "timing belt" stems from the fact that this kind of toothed belt is commonly used in piston engines, connecting the camshaft (used to trigger valve openings), with the crankshaft, in order to "time" valve opening in sync with cylinder compression.

Torque

"Torque", measured in (imperial) lb-in (pounds-inch, "pounds per inch") or (metric) Nm (newton meter) is the rotational equivalent of a linear force ("axial force"), a twisting force applied to an object. In calculations, one has to differentiate between ideal systems and real-world systems. While the torque delivered by a motor (in theory) is constant ("ideal motor"), in reality motors deliver a variable torque, which changes depending on their speed and other factors. And to approximate torque calculations to real-world situations, more variables - like gravity, friction (also proportional to speed, "viscous friction", and air drag - have to be taken into account. When a pulley is attached to the shaft of a motor, the distance between fulcrum of the motor's shaft and the surface of the pulley creates a lever of a certain length, equal with the radius of the lever. With this system, the *force delivered × radius = torque delivered*. Transmission systems are used to multiply or divide torque (cmp. "Gear ratio").

What is the torque required to lift a certain weight?

In order to solve this, you need to know the mass, the weight of the object in *N* ("newtons", cmp "Newton"). The calculation is

*mass * standard gravity = Newtons required*

so in our case the calculation is: *1 * 9.8 = 9.8N* (compare "Standard gravity" to understand why we multiply by 9.8 here). The second step in order to know required motor torque ist to know the length of the lever in effect on the motor's shaft. For example, if the weight is hanging off a rope spun around a pulley, the distance of the pulley surface, where the rope's force is applied and the fulcrum of the motor shaft, where the pulley is attached, is equal to the radius of the pulley. The calculation is

*Force (F) in Netwon * radius in meters = Newton-meters required to hold the mass against gravity.*

So *4,41 × 0,05 = 0.22 Nm*. Moving (accelerating) this mass means we have to add additional force to do so. For example *1 kg × 5 m/s = 5 N*. Adding this to 4.41 N in the previous calculation results in 0.47 Nm required to accelerate the mass at 5m/s. As this doesn't take friction into account, this is just an approximation.

TTL

term in electronics, abbreviation of "Transistor–transistor logic". In microcontroller contexts, speaking of "TTL levels" colloquially refers to the interfacing voltage used on microcontroller inputs/outputs, either being on 3.3V, 5V or 12V levels and often times requiring some form of level shifter to work properly. Logic Levels are a common source of confusion in (DIY) serial communication, especially when RS232 or RS485 interfaces don't adhere to standards.

Robot Types

Over the years, a number of robot categories evolved, each catering to specific needs, having their strengths and weaknesses. Industrial robots are usually selected by the type of work that has to be done, and work might require speed, or strength, have a circular or rectangular work envelope, may require clean-room sealing or dust protection. The robot types usually are:

- Polar Robots (Spherical Robots, Spherical coordinate robots)
- Cylindrical Robots (Cylindrical coordinate robots)
- Articulated Robotic Arms (Revolute coordinate robot)
- Cartesian Robots (Rectangular Robots) (with sub-category Gantry Robots)
- SCARA Robots
- Delta Robots (aka "Parallel Robots")
- Collaborative Robots ("CoBots")

UAV

an "unmanned aerial vehicle" (UAV) is a type of autonomous or remotely piloted robotic vehicle operating without human presence. UAVs are usually teleoperated, sometimes partly autonomous, with autopilot capabilities, to navigate towards set waypoints or to remain airborne in case of loss of a radio connection or while human tele-operators take on other tasks. When flying autonomously, drones utilize predefined software-controlled routes in their embedded systems coupled with various types of sensors and a global positioning system (e.g. GPS). There are two primary types of drone platforms: fixed wing and rotary - each possessing distinctive advantages or disadvantages and thus being chosen based on specific operation requirements. Fixed wing configurations are usually selected for longer missions, in surveillance or for elongated drone presence, as rotary wing setups are usually more taxing on fuel or energy. Drones play a crucial role in accessing remote or hazardous areas, capturing detailed aerial data, performing military operations, delivering supplies to inaccessible areas, etc. Compare ground-based "Unmanned Ground Vehicles" (UGVs)

Uncanny Valley

In an essay, philosopher Walter Benjamin once described that humans react with disgust when they get confronted with things in a state of flux between to states. As an example, he mentions the widespread dislike of the "skin on the milk" that forms when it is boiled. People experience a repulsive impulse that originates in aesthetics that are off, misplaced or slightly wrong, and this impulse has its peak right at the transition from one state to the other. Benjamin suspected that this behavior has its roots in earliest child development, when the body separates from the liquid form of inception, becoming a self-contained being yet retaining the continued fear of losing this embodiment. In 1970, Japanese robotics professor Masahiro Mori found a similar repulsive emotion in people's reaction to robots that looked human-like yet still were "not quite" human-like enough. In one of his books, Mori drew a graph where "human likeness" is plotted on the x-axis and the amount of "familiarity" is plotted on the y-axis. With values of "human likeness" going towards 100%, the curve of "familiarity" is steadily increasing, only to experience a sharp drop right before reaching 100%. This low on the curve he described as the "uncanny valley", where acceptance of a robot, a doll or more general, an object as being "human like" is challenged the most. This refusal is triggered the most when well known features are mixed with synthetic features, like present in some humanoid robot designs or when meeting a person with a prosthetic hand.

Unimate

was the first industrial robot and the first product of the company that produced it, "Unimation, Inc.". Unimation (short for "Universal Automation") was founded in 1956 by US inventors George Devol and Joseph Engelberger. An Unimate prototype was presented in 1959 and was used on a General Motors automobile production line, in 1961 it was working for Ford. In 1968 engineers of Japanese company Kawasaki signed a technical license agreement with Unimation and shortly thereafter, licensed Unimates were domestically built in Japan. Unimation was purchased by Westinghouse in 1983 and later sold to Swiss robotics company Stäubli. Unimate's main arm was mounted on a 2 axis platform, allowing the robot to move in a circular motion and lift or lower the main arm like a crane. This design is called a "Polar Robot", with its arm extending and retracting from the pivotal axis. Unimate's arm was linearly extendable via electronically controlled hydraulics. In total, the robot had six programmable axes and was quite strong, being able to lift 500lbs. Later, Unimation presented the "PUMA" (short for "Programmable Universal Machine for Assembly" or "Programmable Universal Manipulation Arm") articulated robot arm, a robot that resembles more what industrial robots usually look like today, replacing hydraulics with electric actuators on rotational axes. It was available in varying strengths and configurations, as PUMA 500, PUMA 560, PUMA 700, PUMA 200, etc.

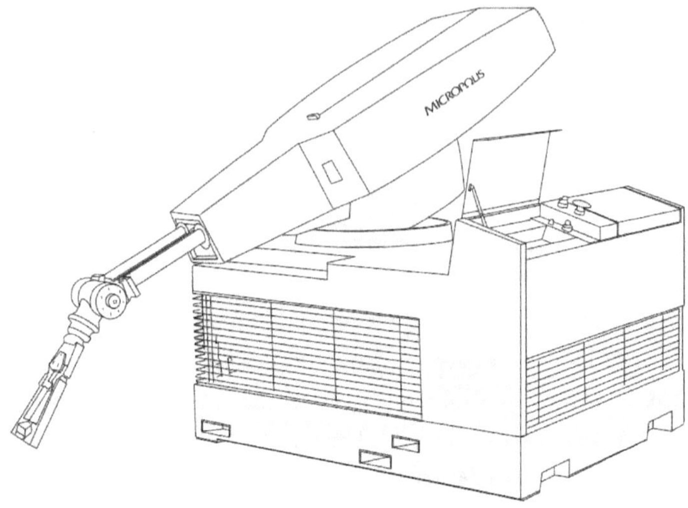

Unimate, the first industrial robot, as introduced in 1959

Related reading:

- There's an in-depth article [spectrum.ieee.org/unimation-robot] about the Unimate at IEEE

- and some more [robotsguide.com/robots/unimate] at the IEEE's robotsguide

- Unimate [en.wikipedia.org/wiki/Unimate] Wikipedia article

USRobotics

often abbreviated as "USR", is a producer of computer modems and despite its name, not active in robotics or industrial automation. The company was a dominant force in the 1990s' BBS and dial-up [www.micropolis.com/support/kb/micropolis-bbs-faq] era. Today (2024), USR, as part of conglomerate UNICOM Global, still manufactures computer modems, being one of the few vendors left.

Via Point

term used in robot programming, defining a point through which the robot's end effector should pass while moving along a programmed path. A via point is passed without stopping and are used helper points to avoid obstacles or optimize the robots posture at a given point in time of a motion cycle.

Work Envelope

The volume defined by the robot's size, degrees of freedom and rotational or linear capabilities of its axes. The physical work envelope of a robot may be restricted to a sub-volume of the work envelope to an "operating envelope" through the use of software and/or limit switches. Compare "Reach".

Worm gear

"Worm gears" are a type of mechanical assembly where two non-intersecting shafts are oriented towards each other at a 90 degree angle (that means they are "cross-axis", or non-parallel). When these two shafts are equipped with gears for a mechanical linkage, and one gear is of a "toothed" design and the other a "screw-like" design, we speak of a worm gear. Worm gears are a simple one-stage gear box design, allowing high torque ratios through high reduction. One common application for worm gear drives is automobile windshield wipers. Worm gears are related to Hypoid gearboxes.

XY-Gantry

A XY-Gantry is a type of positioning device where two axes, the "x" and "y" axis, are manipulated by motors to move one central tool (head) into position on a rectangular working area. XY-Gantries can be described as simple forms of a two-axes robot, a cartesian robot system. The base of the system is a pair of tracks used to move a perpendicular third track in parallel over the work area (Axis 1, "X"). The tool head is mounted on this third track, which is similarly motor controllable to move the head along the third track over the work area (axis 2, "Y"). The tracks are often stepper-motor driven (via belt-systems) with a leadscrew. An XY-Gantry is sometimes called a "cartesian gantry" due to the coordinate system used. A broader term is "linear robot", due to the principal axes of the system being linearly controlled. A combination of terms i "Cartesian coordinate robot". One implementation of a XY-Gantry is CoreXY [en.wikipedia.org/wiki/CoreXY], sometimes "core(X,Y)", which implements the gantry movement via more elaborate pulley systems in order to reduce the moved mass. Another related system are "X-Y tables" (aka "cross working tables" or "coordinate tables", where a flat surface is moved along the motorized axes instead of just a single tool head. Similar systems to XY-Gantries, not working on a flat bed but on a vertical shelf-like work-area are automated warehouses or tape libraries as used in large-scale and enterprise data storage.

Zero Moment Point (ZMP)

is the theoretical ground point where the ground reaction forces counterbalance the inertial forces, causing a state of dynamic equilibrium. In other words, ZMP is a theoretical point on the ground where torque around that point is zero. As ZMP provides the criterion for maintaining bipedal locomotion balance, thus providing stable and controlled movement, ZMP is crucial in the field of humanoid robotics development.

More to read on Robotics & Automation

- The Robotics Primer [pages.ucsd.edu/~ehutchins/cogs8/mataric-primer.pdf] by Maja J. Mataric (Courtesy of UC San Diego); also illustrated from MIT Press, ISBN: 978-0262633543
- The Robotics Primer Workbook [roboticsprimer.sourceforge.net/wiki/index.php/Main_Page]
- "I, Robot" a book by Isaak Asimov, available in numerous editions
- the IEEE robots Guide [robotsguide.com/] is a magazine-like encyclopedia of robots and related tech, from entertainment and industry
- Carnegie Mellon University presents the Robot Hall of Fame [www.robothalloffame.org/inductees.html] (RHOF)
- Northwestern University's Mechatronics Design Wiki [hades.mech.northwestern.edu/index.php/Main_Page]
- "Modern Robotics" by Kevin M. Lynch and Frank C. Park, Cambridge University Press, 2017, ISBN 978-1107156302 [en.wikipedia.org/wiki/Special:BookSources?isbn=9781107156302]; in parts available as Online courses
- "Introduction to AI Robotics" by Robin R. Murphy, ISBN: 9780262133838 [en.wikipedia.org/wiki/Special:BookSources?isbn=9780262133838]
- "Artificial Intelligence: A modern Approach" by Stuart Russell & Peter Norvig, ISBN: 978-1292401133 [en.wikipedia.org/wiki/Special:BookSources?isbn=9781292401133]

Our Handbooks Series

If you enjoyed this book, then we would like to recommend these other publications that have already appeared in the series:

- **Micropolis BBS Primer**
  All the basics of connecting to bulletin board systems, back then and today.
- **Micropolis Data Storage Primer**
  A quick-start guide on basic concepts and advanced topics of computer data storage.

Follow Micropolis online

You can reach Micropolis on our official website and via various social media platforms:

> https://www.micropolis.com/

On YouTube

> https://www.youtube.com/@MicropolisOfficial
> https://www.youtube.com/@MicropolisUsersGroup

On Instagram

> https://www.instagram.com/micropolis.official/

Join the Discussion on Reddit

> https://www.reddit.com/r/Micropolis/

More links and ways to connect on our website.

About Micropolis

Micropolis is a new venture, determined to breathe life into a legendary brand, a computer brand well-known for pioneering electronic and mechanical data storage systems. Under new management, Micropolis strives to live up to the high level of reliability and technical excellence past customers have come to expect, and providing new users with products they can trust on for years.

Brand Legacy

Micropolis was established as a brand in the 1970s. In a time when home-computers began to become a reality for private users. Micropolis Corporation entered the business as a manufacturer of floppy drive controller cards and, building on this early expertise, started to manufacture its own line of 5.25-inch floppy disk drives soon after.

Just as today, storage demands back then grew and customers asked for ever higher data density and reliability for their storage subsystems. Micropolis Corporation answered these demands with the introduction of "rigid disk" systems. Stemming from its experience with controller electronics and mechanical drive systems grew a line of early 8-inch hard-disk drives. In the following years, this product lineup saw miniaturization and higher capacities, with drives offered in the 5.25-inch full-height and later 3.5-inch form factor.

Twenty years after its inception, during the 1990s, the company wasn't only offering hard-disk-drives but had branched out into storage enclosures. A line of deskside "data silo"-type enclosures offered plug-in data storage, packaged in a smart and modular subsystem. But this visionary technology, which anticipated today's storage virtualization, didn't help when many competitors had entered the storage market during the decade. Despite a loyal customer base, Micropolis Corporation experienced cost pressure, was sold, restructured and ultimately, in a market shakeout, ceased operation in 1997.

Micropolis' Future

Today, over 40 years later, the computing landscape has changed dramatically. Data storage density has grown exponentially and systems have seen vast improvements in terms of capabilities and miniaturization. Time has come for a reboot. With a brand like Micropolis, which has such a deep heritage in storage, our contemporary operations will offer customers a line of products that bridge old paradigms with today's demands. Similarly as in Micropolis' first incarnation, new Micropolis will again cater to storage related use-cases and our operations will aim to live up to proven expectations of Legendary Reliability. Legendary Excellence.[SM]

www.ingramcontent.com/pod-product-compliance
Lightning Source LLC
LaVergne TN
LVHW011040200726